Financial Econometrics Using Stata

Financial Econometrics Using Stata

SIMONA BOFFELLI
University of Bergamo (Italy) and Centre for Econometric Analysis, Cass Business School, City University London (UK)

GIOVANNI URGA
Centre for Econometric Analysis, Cass Business School, City University London (UK) and University of Bergamo (Italy)

A Stata Press Publication
StataCorp LP
College Station, Texas

Published by Stata Press, 4905 Lakeway Drive, College Station, Texas 77845
Typeset in LATEX 2ε
Printed in the United States of America

10 9 8 7 6 5 4 3 2 1

Print ISBN-10: 1-59718-214-1
Print ISBN-13: 978-1-59718-214-0
ePub ISBN-10: 1-59718-215-X
ePub ISBN-13: 978-1-59718-215-7
Mobi ISBN-10: 1-59718-216-8
Mobi ISBN-13: 978-1-59718-216-4

Library of Congress Control Number: 2016955738

Contents

Figures

Preface

In this book, we illustrate how to use Stata to perform intermediate and advanced analyses in financial econometrics. The book is mainly for graduate students and practitioners who have an average econometric background. We provide a comprehensive overview of ARMA modeling, as well as univariate and multivariate GARCH models. Our approach consists of presenting a brief but rigorous summary of the theoretical framework, which we then implement using many examples. In particular, we report several empirical applications using real financial markets data to illustrate how to model conditional mean and conditional variance of typical financial time series. Users can easily replicate all the applications, executed using Stata 14, with the datasets and do-files we provide to get familiar with the techniques and Stata commands.

Throughout the book, we use acronyms extensively. For your convenience, we have included a glossary of acronyms at the end of the book.

The book is organized as follows. Chapter 1 provides an introduction to the following: the main features of financial time series, commands for obtaining descriptive statistics, analyzing normality, conducting stationarity tests, autocorrelation, heteroskedasticity, and model selection criteria. Chapter 2 provides a detailed description of the univariate ARMA framework to model the conditional mean of financial time series, with a specific focus on the S&P 500 returns time series.

Chapter 3 introduces the notion of conditional volatility and the popular family of GARCH models, specifically designed to capture the autoregressive nature of the volatility of asset returns. Brief descriptions of GARCH-M, asymmetric GARCH (SAARCH, TGARCH, GJR, APARCH) models, and nonlinear GARCH (PARCH, NGARCH, NGARCHK) models are followed by empirical implementations considering the S&P 500. Chapter 4 extends the univariate GARCH models to the multivariate framework, to account for not only volatility but also correlations between assets. Seminal multivariate GARCH models, such as vech and BEKK models, are described mainly to highlight the curse of dimensional issues; the chapter largely focuses on the CCC and DCC models widely used in the profession. Extensive empirical applications are conducted using four stock indices to stress the empirical validity of the MGARCH framework.

The last two chapters focus on risk management and contagion analyses, two leading research themes among academics and practitioners in the field of financial econometrics. In particular, chapter 5 introduces the concept of risk, risk measures, and their properties, concluding with an overview on some unilevel VaR and multilevel VaR backtesting procedures proposed in the literature. The empirical applications reported illustrate the methods and the way to implement them. Chapter 6 focuses on contagion analysis, where alternative methodologies are presented to evaluate the presence of a contagion. The techniques are illustrated by empirical applications examining the presence of a contagion among the United States, the United Kingdom, Germany, and Japan.

We acknowledge several people to whom we are in debt. First, we are grateful to David Drukker for having sponsored and encouraged us to pursue this project. His support was vital throughout the long gestation of the book, and he read and commented on several drafts of it. Second, we thank Elisabetta Pellini, who carefully read the complete version of the book and provided detailed and constructive feedback at various stages of the project on both the completion of the final document and the empirical applications. Third, we thank Jan Novotny for providing us with useful comments to a preliminary version of the book. Finally, we thank Lisa Gilmore and Deirdre Skaggs for production, LaTeX, and editorial assistance. Any mistakes within the book are ours.

Notation and typography

In this book, we assume that you are somewhat familiar with Stata: you know how to input data, use previously created datasets, create new variables, run regressions, and the like.

We designed this book for you to learn by doing, so we expect you to read it while at a computer trying to use the sequences of commands contained in the book to replicate our results. In this way, you will be able to generalize these sequences to suit your own needs.

We use the `typewriter` font to refer to Stata commands, syntax, and variables. A "dot" prompt followed by a command indicates that you can type verbatim what is displayed after the dot (in context) to replicate the results in the book.

The data we use in this book are freely available for you to download, using a net-aware Stata, from the Stata Press website, http://www.stata-press.com. In fact, when we introduce new datasets, we load them into Stata the same way that you would. For example,

```
. use http://www.stata-press.com/data/feus/spdaily
```

Try it. To download the datasets and do-files to your computer, type the following commands:

```
. net from http://www.stata-press.com/data/feus/
. net describe feus
. net install feus
. net get feus
```

This text complements the material in the Stata manuals but does not replace it, so we often refer to the Stata manuals using [R], [P], etc. For example, [TS] **mgarch ccc** refers to the *Time-Series Reference Manual* entry for `mgarch ccc`, and [P] **syntax** refers to the entry for `syntax` in the *Stata Programming Manual*.

1 Introduction to financial time series

1.1 The object of interest

Financial econometrics deals with estimating the parameters of well-defined probability models that describe the behavior of financial time series, testing hypotheses on how financial markets generate the series of interest, and forecasting future realizations of financial time series (Bollerslev 2001; Engle 2001). The main aim of this book is to describe how we can empirically conduct these analyses using Stata.

The most important financial time series are asset prices, exchange rates, and interest rates. Regarding asset prices, the financial literature deals mainly with returns. Several definitions of returns are used in practice. Considering a time series of closing prices P_t, we can define the following:

- *One-period simple* or *arithmetic returns* are obtained from the basic relationship of financial mathematics $P_t = P_{t-1}(1 + r_t)$ so that returns are given by $r_t = P_t/(P_{t-1} - 1)$.

- *Continuously compounded* or *logarithmic returns* are obtained from $P_t = P_{t-1} \exp(r_t)$, where r is the single-period return. Returns are therefore defined as $r_t = \ln(P_t/P_{t-1})$.

Usually, financial econometricians work with logarithmic returns because they have better properties with respect to arithmetic returns. First, by assuming that returns are normally distributed, prices will by definition follow the log-normal distribution. The log-normal distribution ensures that prices are positive. Second, logarithmic returns are usually adopted to derive closed-form solutions for pricing formulas.

A baseline hypothesis underlying the behavior of the markets is the efficient market hypothesis. We can formulate this market feature as follows: if prices reflect all the expectations and information available to all market participants at time t, then price changes at time $(t + 1)$ must be unpredictable. The concept of efficiency is therefore in strict connection to the availability and accessibility of information; in particular, we can define three forms of efficiency: weak, semistrong, and strong. Starting from the weak form, a market is said to be efficient if past prices could not be used to predict future prices in the long run. From a mathematical point of view, this implies that prices (or returns) do not exhibit any serial dependence that could be modeled by some statistical models. The semistrong form of efficiency supposes that prices adjust very fast to publicly available information. Finally, the strong form of efficiency requires all information, public and private, to be already discounted into the prices in such a way that it will be impossible to make forecasts and beat the market.

In mathematical terms, we can think of a financial time series as the realization of a stochastic process. We now give a formal definition of a stochastic process:

Definition 1.1. A stochastic process $X = \{X\}_{t \geq 0}$ is a collection of random variables, with $t = 1, \ldots, T$ being the time index. Therefore, a stochastic process represents the evolution of random values over time.

According to the nature of the time index t, the stochastic process can be discrete if $X = \{X\}_{t \geq 0}$, $t = 0, 1, 2, \ldots, T$, meaning that time is measured at specific points, or can be continuous if $X = \{X\}_{t \geq 0}$, $0 < t < \infty$, implying that time corresponds to the real line.

We restrict the type of stochastic process that a financial time series must be, because we cannot analyze the properties of an arbitrary stochastic process. In particular, theoretical and empirical research assume that a financial time series can be represented by a stochastic process that has both martingale and Markov properties.

Definition 1.2. A stochastic process $\{X\}_{t \geq 0}$ is said to be a martingale if for any $t \geq 0$ the following properties are satisfied:

1. $E(|X_t|) < \infty$
2. $E(X_{t+1}|X_1, \ldots, X_t) = X_t$

The second property defines the efficient market hypothesis; that is, the expected value of the stochastic process conditioned on the entire available time series is equal to the most recent value of the time series. A financial time series satisfying the efficient market hypothesis is said to follow a martingale process.

Another property satisfied by returns is that of Markovianity.

Definition 1.3. A stochastic process $\{X\}_{t \geq 0}$ is said to be a Markov process if for any $t \geq 0$ the following condition holds true:

$$E\left(X_{t+1}|X_1, \ldots, X_t\right) = E\left(X_{t+1}|X_t\right)$$

This definition implies that the process is memoryless, which means that the expected value conditioned on the entire information set is equal to the expected value conditioned on the most recent observation.

1.2   Approaching the dataset

We briefly describe the data that we use throughout the first three chapters to introduce the commands of interest via some empirical applications. We choose the S&P 500 closing prices over the years 1950–2013 at daily frequency for a total of 16,103 observations.

On closing prices, we compute logarithmic returns. The data were initially available from an Excel source. In the appendix of this chapter, we discuss how to import these data from Excel.

The first step when analyzing a time series is to define a variable accounting for the time. In particular, we must use the `tsset` command to identify this variable. For instance, in our case, we type

```
. use http://www.stata-press.com/data/feus/spdaily
. tsset newdate
```

Once we have `tsset` our dataset, we are ready to create the first chart for prices and returns by using the `tsline` command. `tsline` offers several options to customize the format of the chart; for instance, we may specify the format for the markers ([G-3] *marker_options*) or add labels for the markers ([G-3] *marker_label_options*). `tsline` does not require specification of the variable to use for labeling the *x* axis because Stata already knows this information from the `tsset` command.

We start by drawing the chart of the prices time series by typing

```
. tsline price
```

which produces figure 1.1:

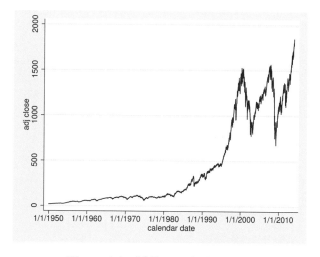

Figure 1.1. S&P 500 daily prices

To show some options of `tsline`, we now draw the time series of returns, where we add the label "Black Monday" by using the following command. We have used the `ttick()` and `ttext()` options to point out the Black Monday on the time axis, as depicted in figure 1.2.

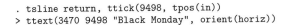

```
. tsline return, ttick(9498, tpos(in))
> ttext(3470 9498 "Black Monday", orient(horiz))
```

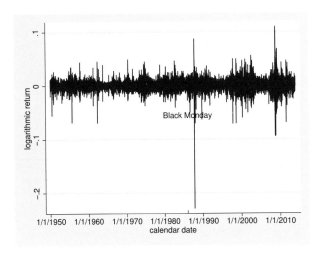

Figure 1.2. S&P 500 daily returns

We now use the `describe` command to get a description of the dataset currently loaded.

```
. describe
Contains data from http://www.stata-press.com/data/feus/spdaily.dta
  obs:        16,103
  vars:           11                          6 Feb 2016 16:41
  size:    1,062,798
```

	storage	display	value	
variable name	type	format	label	variable label
date	int	%td..	.	calendar date
open	double	%10.0g		open price
high	double	%10.0g		high price
low	double	%10.0g		low price
close	double	%10.0g		close price
volume	double	%10.0g		traded volume
price	double	%10.0g		adj close
newdate	long	%tbspdaily		business date
logprice	float	%9.0g		logarithmic price
return	float	%9.0g		logarithmic return
blackmonday	float	%9.0g		

```
Sorted by: date
```

The output provides us with some basic information about the dataset. There are 16,103 observations on 11 variables. The table provides the variable names, storage types, display formats, value labels, and variable labels. `blackmonday`, `logprice`, and `return` are stored as `float`, `date` as `int`, `newdate` as `long`, and the other variables as `double`. See [D] **data types** for the available storage types and the space–range trade-off among them. The display format determines how observations are displayed when they are listed; see [D] **format** for details. Value labels, not specified here, are strings that explain numeric codes; see [D] **label** for details. The variable label is a short string that describes the variable. The footer tells us that the dataset is sorted by `date`.

Once we have information about the variables in the dataset, it is useful to get some insight on the statistical properties characterizing the variables. For this, we use the `summarize` command:

```
. summarize return, detail
```

```
                        logarithmic return
```

	Percentiles	Smallest		
1%	-.0261493	-.2289972		
5%	-.014503	-.0946951		
10%	-.0099645	-.0935364	Obs	16,102
25%	-.004128	-.0921898	Sum of Wgt.	16,102
50%	.000468		Mean	.0002925
		Largest	Std. Dev.	.0097578
75%	.0049715	.0683665		
90%	.0101552	.0870891	Variance	.0000952
95%	.0144796	.102457	Skewness	-1.030111
99%	.0255079	.1095724	Kurtosis	30.69293

The right-hand side of the table reports the estimated mean, standard deviation, variance, skewness, and kurtosis coefficients. In particular, we get skewness and kurtosis coefficients by adding the `detail` option to `summarize`. Recall that skewness measures the symmetry of the distribution: left-skewed distributions have a negative skewness coefficient, symmetric distributions have a skewness coefficient of 0, and right-skewed distributions have a positive skewness coefficient. In our case, returns are left-skewed, indicating the presence of a long left tail, probably caused by the presence of some negative outliers. This is a common feature characterizing returns distribution. Formally, skewness $S(x)$ is the normalized third moment of a random variable X:

$$S(x) = E\left\{ \frac{(X - \mu_x)^3}{\sigma_x^3} \right\}$$

with μ_x being the first moment and σ_x being the second central moment of X. The sample counterpart of $S(x)$ is given as

$$\widehat{S}(x) = \frac{1}{(T-1)\,\widehat{\sigma}_x^3} \sum_{t=1}^{T} (x_t - \widehat{\mu}_x)^3$$

where $\widehat{\mu}_x$ and $\widehat{\sigma}_x$ are the sample mean and the standard deviation, respectively.

The kurtosis coefficient measures the heaviness of the tails of the distribution. Heavy-tailed distributions produce more extreme values for a given mean and variance. The benchmark distribution in financial econometrics is the normal distribution, which is characterized by a kurtosis coefficient equal to 3. In our case, the S&P shows a kurtosis coefficient equal to 30.69, indicating the presence of heavy tails and suggesting the presence of more extreme values than we would expect from a normal distribution. Again, the presence of heavy tails is a typical feature characterizing returns distributions. Formally, kurtosis $K(x)$ is the normalized fourth moment of a random variable X:

$$K(x) = E\left\{ \frac{(X - \mu_x)^4}{\sigma_x^4} \right\}$$

with μ_x and σ_x being the first moment and the second central moment of X, respectively. The sample counterpart of $K(x)$ is given as

$$\widehat{K}(x) = \frac{1}{(T-1)\,\widehat{\sigma}_x^4} \sum_{t=1}^{T} (x_t - \widehat{\mu}_x)^4$$

where $\widehat{\mu}_x$ and $\widehat{\sigma}_x$ are the sample mean and the standard deviation, respectively.

The first four moments of a return distribution characterize the behavior of a financial asset. The first moment is the mean, and it indicates whether the asset tends to generate positive or negative returns. The second moment is the variance, and it measures the uncertainty, risk, and volatility of an asset. As the second moment gets larger, the return distribution becomes more widely dispersed, and we become less certain about possible realized values. This loss of certainty implies that decisions based on returns become more risky. This loss of certainty is also known as increased volatility.

The third moment is the skewness, and for unimodal continuous distributions, it measures the relative frequency of left-tail realizations to right-tail realizations. When we are working with a unimodal distribution, a negative skew could indicate that the left tail is longer or fatter than the right tail. On the other hand, when the skew gets positive, the distribution is characterized by a longer or fatter right tail. The interpretation of the skewness is not as simple when one of the two tails is longer but the other is fatter. In this case, the skewness averages out the two indications, and we cannot summarize in a number the shape of the distribution. This issue gets even more complicated when we are dealing with multimodal distributions or discrete distributions. When working with returns of financial assets, a negative skewness indicates that the asset will tend to generate big negative returns, and therefore it is riskier. Vice versa, a positive skewness suggests that the asset will tend to generate big positive returns.

The fourth moment is the kurtosis, and it measures the frequency of tail realization relative to a central realization. A return distribution with higher kurtosis generates extreme returns, both positive and negative, more frequently. Therefore, an asset characterized by higher kurtosis will be riskier compared with an asset that tends to generate fewer extreme returns. A high variance, a high negative skewness, and a high kurtosis all indicate a riskier asset.

The third column in the table reports the four smallest and the four largest returns that, as we can see, are quite far from the 1st and the 99th percentiles. We note that the minimum value is -22.9%, which we know to correspond to the Black Monday on 19 October 1987. This single observation can greatly explain on its own the high value for both kurtosis and skewness.

We now analyze the magnitude of Black Monday. To this purpose, we re-examine the returns series while excluding that observation by adding the condition `if newdate != d(19Oct1987)`.

```
. summarize return if date != td(19Oct1987), detail

                         logarithmic return

              Percentiles      Smallest
       1%       -.0260973      -.0946951
       5%        -.014503      -.0935364
      10%       -.0099645      -.0921898     Obs               16,101
      25%       -.0041265      -.0864186     Sum of Wgt.       16,101

      50%        .0004683                    Mean            .0003067
                                Largest      Std. Dev.       .0095893
      75%        .0049715       .0683665
      90%        .0101552       .0870891     Variance         .000092
      95%        .0144796        .102457     Skewness       -.2407553
      99%        .0255079       .1095724     Kurtosis        12.60677
```

Alternatively, we could have exploited the dummy variable `blackmonday` by specifying the condition `if blackmonday==0`.

The results show that the kurtosis is still larger than 3, confirming that the distribution is still far from being normal. The kurtosis is now equal to 12.61 instead of the 30.69 we got when we were not controlling for Black Monday. This difference provides some clear evidence of the impact that a single outlier can have on the kurtosis. In addition, the removal of the observation on Black Monday has a great impact on skewness as well, which decreases from -1.03 to -0.24 and therefore becomes less skewed.

There are interesting links between the sampling frequency of financial time series (the frequency at which the time series is recorded) and estimated moments. For instance, as the sampling frequency increases, moving from monthly to daily, the kurtosis is likely to increase as well because it is more likely to observe extreme returns in the daily distribution than in the monthly distribution. To appreciate the strength of the relationship between resampling frequency and the statistical indicators, we now consider monthly returns of the S&P 500 index, and we report the descriptive statistics of interest.

```
. use http://www.stata-press.com/data/feus/spmonthly, clear
. summarize return, detail
```

```
                              logarithmic return

              Percentiles        Smallest
     1%       -.1073568         -.245428
     5%       -.0641109         -.1856365
    10%         -.04653         -.1575861      Obs                   768
    25%       -.0178209         -.1270777      Sum of Wgt.           768

    50%        .0092016                        Mean            .0060599
                                 Largest       Std. Dev.       .0420061
    75%        .0346221          .1118149
    90%        .0526409          .1158364      Variance        .0017645
    95%        .0687726          .1237801      Skewness       -.6630392
    99%        .1023066          .1510432      Kurtosis         5.42059
```

The kurtosis is equal to 5.42 at the monthly frequency, whereas it was equal to 30.69 when we were using daily data. The intuition underlying this empirical evidence relies on the smoother distribution characterizing monthly returns compared with daily returns. In fact, when computing monthly returns, we are averaging out daily movements, which are more subject to shocks or bad news.

1.3   Normality

Most econometric procedures are derived under the assumption of Gaussian distribution thanks to its statistical properties and analytic tractability. However, returns series are widely recognized not to satisfy the normality assumption. Therefore, we now introduce some testing procedures to evaluate the normality of returns distribution.

A good starting point is a visual inspection of the plot of the time series. The first
chart we report is the basic histogram (see figure 1.3), which we can obtain using the
`histogram` command. For instance, we can evaluate the normality of S&P 500 returns
as follows:

```
. use http://www.stata-press.com/data/feus/spdaily
. histogram return, normal normopts(lcolor(gs13)) kdensity
> kdenopts(lcolor(black))
(bin=42, start=-.22899723, width=.00806118)
```

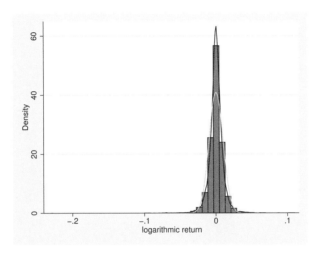

Figure 1.3. Histogram of daily S&P returns

We add the kernel distribution (in black in figure 1.3) by specifying the `kdensity`
option, and we add the normal distribution by specifying the `normal` option. We customize the normal distribution format by specifying the `normopts()` option; in this
case, we changed the line identifying the normal distribution to be light gray. The normal distribution reported on the chart has the same location and scale parameters as
the empirical distribution. As per the kernel distribution, the estimates are produced
by default using the Epanechnikov kernel with an optimal half-width; we can select
an alternative kernel by using the option `kdenopts()`. See [R] **histogram** for further
details.

Figure 1.3 indicates that the empirical distribution is far from normal, being more
fat-tailed and more peaked than the normal distribution would require.

To evaluate the impact of Black Monday, we drop this observation from our sample.
In addition, we increase the number of bins to obtain a more detailed chart by using
the `bin()` option. Figure 1.4 shows that the S&P returns without Black Monday still
depart from the normality assumption, although to a lesser extent with respect to the
full sample.

```
. histogram return if newdate != td(19Oct1987), bin(100) normal
> normopts(lcolor(gs13)) kdensity kdenopts(lcolor(black))
(bin=100, start=-.22899723, width=.0033857)
```

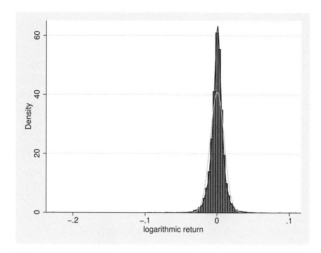

Figure 1.4. Histogram of daily S&P returns without Black Monday

In figure 1.5, we report the distribution of daily returns on the left and of monthly returns on the right.

```
. histogram return, name(HistDaily)
(bin=42, start=-.22899723, width=.00806118)
. use spmonthly, clear
. histogram return, name(HistMonthly)
(bin=27, start=-.24542805, width=.01468412)
```

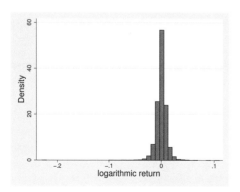 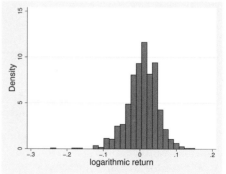

Figure 1.5. Histograms of daily (left) and monthly (right) S&P returns

The two distributions are quite different. The left tail of the daily distribution is much longer than that of the monthly distribution because of the presence in the daily distribution of Black Monday. In addition, the daily distribution is much more peaked than the monthly one, which looks closer to the normal distribution. As already mentioned, this is a common result in financial time series: the higher the frequency at which data are sampled, the higher the departure from normality.

Another tool to assess the normality of a distribution is a more sophisticated and quite common chart called the quantile–quantile plot or Q–Q plot. This chart is based on the comparison between the theoretical quantiles coming from a prespecified distribution—in our case, the normal distribution—and those obtained from the empirical distribution. We can draw the Q–Q plot (see figure 1.6) by using the `qnorm` command.

```
. use http://www.stata-press.com/data/feus/spdaily
. qnorm return, grid
```

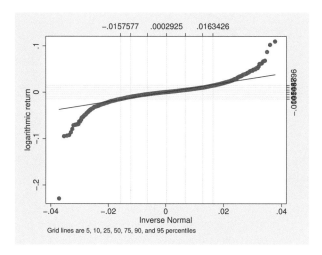

Figure 1.6. Q–Q plot for daily S&P returns

We specified the `grid` option to add grid lines to the chart corresponding to the 5th, 10th, 25th, 50th, 75th, 90th, and 95th percentiles. In addition to the curve comparing the theoretical and empirical quantiles, the chart displays a straight 45-degree line indicating the case of perfect matching between the theoretical and empirical distributions, which is the case when all the quantiles from the two distributions are identical. We have to interpret any deviations from the straight line as a sign of departure from the theoretical distribution. In our case, there is evidence of a significant distance between the curve and the straight line, especially in the tails of the distribution, suggesting that our empirical distribution is far from being normal. Moreover, in the lower part of the chart, we can see the point corresponding to Black Monday, which greatly contributes in our rejection of the hypothesis of normality.

In addition to the Q–Q plot, we can also implement the probability plot or P–P plot, which relies on the cumulative distribution function (c.d.f.) of the two distributions. Thus, while the Q–Q plot refers mainly to the tails of the distribution, the P–P plot focuses on the center of the distribution. The P–P plot is available using the `pnorm` command; see figure 1.7.

```
. pnorm return, grid
```

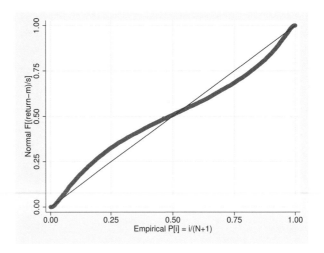

Figure 1.7. P–P plot for daily S&P returns

The vertical axis reports the label `F[(return-m)/s]`, where m and s denote the empirical mean and standard deviation, respectively. On the horizontal axis, N indicates the length of the time series and i indicates the ith observation. Even the P–P plot displays a straight 45-degree line, indicating the case of perfect overlapping between the theoretical and empirical distributions. In our case, the P–P plot confirms that the empirical distribution is far from being normal. Because the P–P plot is more focused on the center of the distribution rather than on the tails, the departure from the normality assumption is less evident when compared with the Q–Q plot.

We now introduce formal testing procedures for assessing the normality of a time series. We start by presenting a test based on skewness and kurtosis, implemented using the `sktest` command. This test builds on the idea of the well-known Bera–Jarque test, but it is based on the asymptotic standard errors with no correction for sample size. We execute the `sktest` on the S&P 500 returns:

```
. sktest return
                    Skewness/Kurtosis tests for Normality
                                                        ─────── joint ───────
        Variable |     Obs   Pr(Skewness)   Pr(Kurtosis) adj chi2(2)   Prob>chi2
    -------------+---------------------------------------------------------------
          return |  16,102        0.0000         0.0000        .              .
```

The sktest command implements three tests. The first, Pr(Skewness), tests the null hypothesis that the skewness is equal to 0; the second, Pr(Kurtosis), tests whether the kurtosis is different from 3; and the third is a joint test for the skewness and the kurtosis for which both the test statistic (adj chi2(2)) and the p-value (Prob>chi2) are reported. In our case, the value of χ^2 is not displayed; this occurs when the value is too high, corresponding to an almost certain nonnormality of the data.

By specifying the noadjust option, we obtain the unaltered test as described by D'Agostino, Belanger, and D'Agostino (1990), which provides us with the value of the test statistic.

```
. sktest return, noadjust
                    Skewness/Kurtosis tests for Normality
                                                        ─────── joint ───────
        Variable |     Obs   Pr(Skewness)   Pr(Kurtosis)     chi2(2)   Prob>chi2
    -------------+---------------------------------------------------------------
          return |  16,102        0.0000         0.0000     6436.73        0.0000
```

The test statistic equals 6,436.73, allowing us to strongly reject the null hypothesis that returns are normally distributed.

We have at our disposal two additional tests for evaluating normality, namely, the Shapiro–Wilk and the Shapiro–Francia normality tests. The Shapiro–Wilk test is implemented with the swilk command. We apply this test on the S&P returns:

```
. swilk return
                    Shapiro-Wilk W test for normal data
        Variable |     Obs        W          V         z      Prob>z
    -------------+------------------------------------------------------
          return |  16,102    0.89741    766.139    17.996    0.00000
```

The Shapiro–Wilk test strongly rejects the null hypothesis that returns are normally distributed. The value reported under W is the Shapiro–Wilk test statistic. The value reported under V is an index accounting for the degree of departure from normality. It should be interpreted as follows: if the variable is normally distributed, then V is close to 1; the more the variable deviates from the normal distribution, the larger V gets. The 95% confidence interval for V is [1.2, 2.4]. In our case, V is equal to 766.14, suggesting that returns are not normally distributed.

The Shapiro–Francia test is implemented by the `sfrancia` command:

```
. sfrancia return
                    Shapiro-Francia W´ test for normal data
      Variable │      Obs        W´         V´          z       Prob>z
      ─────────┼──────────────────────────────────────────────────────
        return │   16,102     0.89665   902.543     18.495     0.00001
      Note: The normal approximation to the sampling distribution of W´
            is valid for 10<=n<=5000 under the log transformation.
```

The Shapiro–Francia test also rejects the hypothesis of Gaussian distribution for the returns. Similarly to the Shapiro–Wilk test, we are provided with a synthetic index for the degree of departure from normality, `V'`, whose 95% confidence interval for accepting the null hypothesis of normality is $[2.0, 2.8]$. In our case, `V'` equals 902.54, clearly lying outside the confidence bounds.

Note that the Shapiro–Wilk test is accurate only when the number of observations lies between 4 and 2,000; Shapiro–Francia is accurate for 5 to 10,000 observations.

To summarize, in this section, we have presented several alternatives to test for normality. Each of the alternatives supplied evidence of departure from normality when applied to the S&P 500 returns series.

1.4   Stationarity

Financial econometricians generally work with returns rather than prices. In general, returns are characterized by time-invariant distribution, meaning that returns follow a stationary process.

Definition 1.4. A time series $\{r\}_t$ is strictly stationary if the joint distribution of $(r_{t_1}, \ldots, r_{t_k})$ is identical to that of $(r_{t_1+\tau}, \ldots, r_{t_k+\tau})$ for all positive integers τ.

Strict stationarity requires that the joint distribution of the subsequence $(r_{t_1}, \ldots, r_{t_k})$ does not change when it is shifted by an arbitrary amount τ. If we consider that stationarity requires that all moments of the joint distribution are invariant to time shifts, we can easily understand that the distributions that generate most financial time series are not strictly stationary.

Thus, we use a weaker definition of stationarity.

Definition 1.5. A time series $\{r\}_t$ is said to be weakly or covariance stationary if the following conditions hold true:

1. $E(r_t) = \mu$: the mean of the process is constant through time and equal to a constant μ;

2. $\text{Var}(r_t) = \gamma_0$: the variance of the process is time invariant and equal to a finite constant $\gamma_0 < \infty$;

3. $\text{Cov}(r_t, r_{t+l}) = \gamma_l$, $|\gamma_l| < \infty$: the covariance of the process should not be time dependent, but it can be affected just by the distance between the two time ticks considered, equal to l.

Therefore, the weak stationarity imposes constraints on just the first two moments of the distribution, while the strict stationarity checks that the entire distribution is time invariant. Thus, weak stationarity does not imply strict stationarity, because the weak stationarity does not impose conditions on moments higher than the second. Nor does strict stationarity imply weak stationarity, because the definition of strict stationarity does not require the variance to be finite. However, under the Gaussian assumption, weak stationarity always implies strict stationarity, because the Gaussian distribution is entirely characterized by its first two moments.

A well-known stationary process is the white-noise process.

Definition 1.6. A return time series $\{r\}_t$ is said to follow an independent white-noise process if it satisfies the following conditions:

1. $E(r_t) = 0$
2. $E(r_t^2) = \sigma^2 < \infty$
3. $E(r_t, r_{t-j}) = 0 \ \forall j \neq 0$

A white-noise process has finite mean and variance, and it does not show any time pattern, meaning that the current realizations of a process cannot help in predicting its future realizations. Therefore, because independence implies absence of autocorrelation, a white-noise process is characterized by almost flat autocorrelation function (ACF) and partial autocorrelation function (PACF), with no correlation statistically different from 0. Returns can usually be ascribed to the class of white-noise process, coherently with the assumption of efficient market hypothesis.

We now simulate a Gaussian white-noise process. Note that normality is not a general requirement for this process. We start by setting the length of our simulation period equal to 1,000 by using the **set obs** command, and we generate a time index (**index**) of the same length. In addition, we set the seed (the starting point for any random sequence) to ensure we get the same sequence of random numbers every time the simulation is run—which is important when we are ready to replicate the simulation. Finally, we extract simulated numbers from a standard normal distribution by using the **rnormal()** function, taking as an argument the mean and the standard deviation that, in our case, we respectively set equal to 0 and to 1.

```
. clear
. set obs 1000
number of observations (_N) was 0, now 1,000
. generate index = _n
. * fix seed
. set seed 1
```

```
. generate wn1 = rnormal(0,1)
. generate wn2 = rnormal(0,1)
. generate wn3 = rnormal(0,1)
. tsset index
        time variable:  index, 1 to 1000
                delta:  1 unit
```

We then use the **tsline** command to graph the results; see figure 1.8.

```
. tsline wn1 wn2 wn3
```

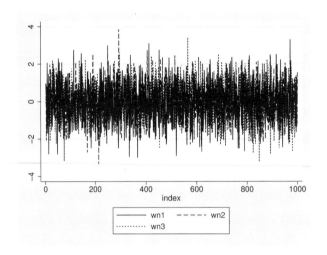

Figure 1.8. Simulated white-noise processes

Although the three processes are almost not distinguishable, they all move around the zero line, suggesting that they are stationary.

A common nonstationary process is the random walk.

Definition 1.7. A time series $\{p_t\}$ is a random walk if it satisfies

$$p_t = p_{t-1} + \varepsilon_t \tag{1.1}$$

where ε_t is a white-noise process.

A random walk is the typical process that is able to describe the behavior of stock prices.

A generalization of (1.1) is the random walk with drift:

$$p_t = \mu + p_{t-1} + \varepsilon_t$$

where μ, commonly called drift, represents the time trend of the log price.

We can obtain a random-walk process (see figure 1.9) as the cumulative sum of the white-noise processes just simulated above.

```
. generate rw1 = sum(wn1)
. generate rw2 = sum(wn2)
. generate rw3 = sum(wn3)
. tsline rw1 rw2 rw3
```

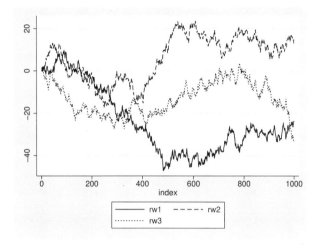

Figure 1.9. Simulated random-walk processes

All three simulated processes show a trend suggesting that they are not stationary.

1.4.1 Stationarity tests

With the purpose of establishing whether a time series is stationary or nonstationary, we can use the unit-root test. A process with a unit root has time-dependent variance, thus violating the condition of weak stationarity, $\text{Var}(r_t) = \gamma_0$.

Stata can test for the presence of a unit root by using two main testing procedures: the augmented Dickey–Fuller (ADF, 1979) test and the Phillips and Perron (PP, 1988) test.

Given a time series $\{y_t\}$, the ADF test is based on the regression

$$\Delta y_t = \alpha + \beta t + \theta y_{t-1} + \delta_1 \Delta y_{t-1} + \cdots + \delta_{p-1} \Delta y_{t-p+1} + \varepsilon_t \tag{1.2}$$

where α is a constant, t is the time trend, and p is the order of the autoregressive process.

The null hypothesis under which the ADF test is distributed is that the time series is not stationary, corresponding to $\theta = 0$, against the alternative that it is stationary, corresponding to $\theta < 0$. The underlying idea is that if the series $\{y_t\}$ is stationary, then the only relevant information to explain Δy_t comes from its lags and not from the lagged value y_{t-1}. The limiting distribution of the test statistic is not standard, but it is known and its critical values have been tabulated.

We can easily carry out the ADF test by using the `dfuller` command. For instance, to test for stationarity of S&P daily log-prices, we type

```
. use http://www.stata-press.com/data/feus/spdaily, clear
. dfuller logprice, lags(2) regress
Augmented Dickey-Fuller test for unit root          Number of obs   =      16100
```

	Test Statistic	1% Critical Value	Interpolated Dickey-Fuller 5% Critical Value	10% Critical Value
Z(t)	-0.866	-3.430	-2.860	-2.570

MacKinnon approximate p-value for Z(t) = 0.7991

D.logprice	Coef.	Std. Err.	t	P>|t|	[95% Conf. Interval]
logprice					
L1.	-.0000509	.0000588	-0.87	0.387	-.0001661 .0000643
LD.	.0296736	.0078746	3.77	0.000	.0142384 .0451087
L2D.	-.0429055	.0078743	-5.45	0.000	-.0583401 -.027471
_cons	.0005656	.0003215	1.76	0.079	-.0000645 .0011957

`dfuller` is followed by the name of the variable we want to test for the presence of a unit root, `logprice`, and by the number of lagged differences to be included in the regression, `lags(2)`. We specify the `regress` option to obtain the regression table, which is omitted by default. Note that we could have obtained in the regression the trend βt in (1.2) by specifying the `trend` option and the drift by specifying the `drift` option.

At the first lag, the test statistic is greater than the critical values, so we cannot reject the null hypothesis that the time series is not stationary.

The nonstationarity of the log-prices is evident even from a visual inspection of figure 1.10.

```
. tsset date
        time variable:  date, 1/3/1950 to 12/31/2013, but with gaps
                delta:  1 day
. tsline logprice
```

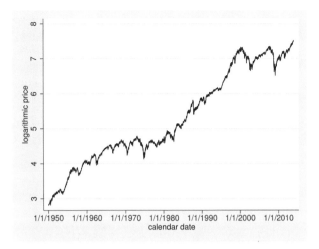

Figure 1.10. S&P daily log-prices

We now evaluate the stationarity of log-returns:

```
. tsset newdate
        time variable:  newdate, 03jan1950 to 31dec2013
                delta:  1 day

. dfuller return, lags(2) regress

Augmented Dickey-Fuller test for unit root         Number of obs   =      16099

                                          ─────── Interpolated Dickey-Fuller ───────
                       Test           1% Critical          5% Critical         10% Critical
                    Statistic            Value               Value                Value

Z(t)                 -74.382            -3.430               -2.860               -2.570

MacKinnon approximate p-value for Z(t) = 0.0000
```

D.return	Coef.	Std. Err.	t	P>|t|	[95% Conf. Interval]	
return						
L1.	-1.009731	.013575	-74.38	0.000	-1.03634	-.983123
LD.	.0395432	.0109775	3.60	0.000	.0180262	.0610602
L2D.	-.0034916	.0078819	-0.44	0.658	-.0189411	.0119578
_cons	.0002942	.0000769	3.82	0.000	.0001434	.0004449

The test statistic is equal to -74.38; therefore, we can reject the null hypothesis of a unit root in favor of the alternative hypothesis that the time series of log-returns is stationary. We also confirm that log-prices are integrated of order 1.

In addition to the classical ADF test, we can use the modified Dickey–Fuller t test, usually known as the DF-GLS test proposed by Elliott, Rothenberg, and Stock (1996). The DF-GLS test has significantly higher power than the ADF test for a small sample.

From a technical point of view, the DF-GLS test differs from the ADF test in that the time series is transformed via a generalized least-squares (GLS) regression before performing the test. Therefore, although the DF-GLS test is based on (1.2)—as is the ADF test— here we consider GLS-detrended data. The null hypothesis is that y_t is a random walk, while the alternative is twofold: the time series is stationary around a linear trend or has a nonzero mean with no linear time trend.

We now evaluate the stationarity of our returns time series by applying the DF-GLS test with the `dfgls` command.

```
. dfgls return

DF-GLS for return                                    Number of obs = 16059
Maxlag = 42 chosen by Schwert criterion

              DF-GLS tau      1% Critical     5% Critical    10% Critical
     [lags]  Test Statistic     Value           Value           Value
     ─────────────────────────────────────────────────────────────────
       42        -5.023         -3.480          -2.836          -2.548
       41        -5.046         -3.480          -2.836          -2.548
       40        -5.141         -3.480          -2.836          -2.548
       39        -5.307         -3.480          -2.836          -2.548
       38        -5.379         -3.480          -2.836          -2.549
       37        -5.554         -3.480          -2.836          -2.549
       36        -5.667         -3.480          -2.836          -2.549
       35        -5.793         -3.480          -2.836          -2.549
       34        -5.937         -3.480          -2.837          -2.549
       33        -6.125         -3.480          -2.837          -2.549
       32        -6.071         -3.480          -2.837          -2.549
       31        -6.325         -3.480          -2.837          -2.549
       30        -6.612         -3.480          -2.837          -2.549
       29        -6.758         -3.480          -2.837          -2.549
       28        -7.019         -3.480          -2.837          -2.549
       27        -7.434         -3.480          -2.837          -2.549
       26        -7.677         -3.480          -2.837          -2.549
       25        -8.125         -3.480          -2.837          -2.549
       24        -8.267         -3.480          -2.837          -2.550
       23        -8.502         -3.480          -2.837          -2.550
       22        -8.921         -3.480          -2.838          -2.550
       21        -9.290         -3.480          -2.838          -2.550
       20        -9.693         -3.480          -2.838          -2.550
       19        -9.926         -3.480          -2.838          -2.550
       18       -10.484         -3.480          -2.838          -2.550
       17       -11.041         -3.480          -2.838          -2.550
       16       -11.366         -3.480          -2.838          -2.550
       15       -11.903         -3.480          -2.838          -2.550
       14       -13.020         -3.480          -2.838          -2.550
       13       -13.707         -3.480          -2.838          -2.550
       12       -14.621         -3.480          -2.838          -2.550
       11       -15.665         -3.480          -2.838          -2.550
       10       -17.431         -3.480          -2.838          -2.551
        9       -18.707         -3.480          -2.839          -2.551
        8       -20.836         -3.480          -2.839          -2.551
        7       -22.920         -3.480          -2.839          -2.551
        6       -26.067         -3.480          -2.839          -2.551
        5       -29.184         -3.480          -2.839          -2.551
        4       -33.778         -3.480          -2.839          -2.551
```

```
     3            -39.774          -3.480          -2.839        -2.551
     2            -49.059          -3.480          -2.839        -2.551
     1            -65.819          -3.480          -2.839        -2.551

  Opt Lag (Ng-Perron seq t) = 41 with RMSE  .0098189
  Min SC   = -9.221803 at lag 40 with RMSE  .0098207
  Min MAIC =  -9.22966 at lag 41 with RMSE  .0098189
```

dfgls evaluates 42 lags for regression in (1.2), and so we obtain very long output.
The first column reports the number of lags included at each step; the second column
shows the calculated statistics; and the remaining three columns report the critical
values at 1%, 5%, and 10% confidence levels, respectively. The null hypothesis of a
unit root is strongly rejected for all the lags considered, confirming that the returns are
stationary. At the bottom of the output of dfgls, we find a description of the optimal
number of lags to be included in (1.2), computed according to alternative criteria: the
Ng–Perron sequential (Ng-Perron seq t), the Schwarz (SC), and the modified Akaike
information criteria (MAIC).

As an alternative to the ADF test, we can use the PP test implemented after the
command dfgls. The PP test is built on the ADF test, and it is based on the regression
in (1.2), where the null hypothesis is that $\theta = 0$. Just as in the ADF test, the PP test
addresses the issue that the process generating data for y_t might have a higher order of
autocorrelation than it is admitted in the test equation. To do that, Stata introduces a
nonparametric correction for the t-test statistic. Thus, the test is robust to unspecified
autocorrelation and heteroskedasticity in the disturbance process of the test equation.
The null hypothesis is that the time series is nonstationary against the alternative that it
is stationary or it does not contain a unit root. The main difference between the PP and
the ADF tests is that the former uses Newey and West (1987) standard errors to account
for serial correlation, whereas the latter uses additional lags of the first difference.

We can run the PP test by using the pperron command.

```
. pperron return
Phillips-Perron test for unit root              Number of obs   =      16101
                                                Newey-West lags =         12

                                     ─── Interpolated Dickey-Fuller ───
                    Test         1% Critical      5% Critical     10% Critical
                 Statistic          Value            Value           Value

  Z(rho)        -14834.209        -20.700          -14.100         -11.300
  Z(t)           -123.310          -3.430           -2.860          -2.570

MacKinnon approximate p-value for Z(t) = 0.0000
```

The PP test confirms that the time series of log-returns is stationary, which leads us
to reject the null hypothesis that the series contains a unit root.

1.5   Autocorrelation

Considering a time series $\{r\}_t$, the autocorrelation characterizes the dependency between r_t and its past values r_{t-k}, $k > 0$. Autocorrelation is therefore the correlation of a process at time t and its past values.

To explore the autocorrelation property of a time series, we now analyze its autocorrelation (AC) and partial autocorrelation (PAC) functions.

1.5.1   ACF

For a weak stationary process, the covariance between current and past values is a function of the distance between the two time realizations, k.

$$\rho_k = \frac{\text{Cov}\,(r_t, r_{t-k})}{\sqrt{\text{Var}\,(r_t)}\sqrt{\text{Var}\,(r_{t-k})}} = \frac{\text{Cov}\,(r_t, r_{t-k})}{\sqrt{\text{Var}\,(r_t)}\sqrt{\text{Var}\,(r_t)}} = \frac{\text{Cov}\,(r_t, r_{t-k})}{\text{Var}\,(r_t)} \tag{1.3}$$

where the equality in the denominators follows from the stationarity assumption of r_t, implying that the variance is equal at each time step. In chapter 2, we will use the ACF to get some preliminary insights about the time structure of the process and to define the best model specification.

The estimator of (1.3) at a generic lag k is given by

$$\widehat{\rho}_k = \frac{\sum_{t=k+1}^{T} (r_t - \bar{r})\,(r_{t-k} - \bar{r})}{\sum_{t=1}^{T} (r_t - \bar{r})^2}$$

If $\{r\}_t$ is an independent and identically distributed (i.i.d.) sequence satisfying $E(r^2) < \infty$, then $\widehat{\rho}_k$ is asymptotically normally distributed with mean 0 and variance $1/T$ for any fixed positive integer k.

Once we have obtained the empirical ACF, the next step is to determine whether the time series exhibits some statistically significant time dependence structure. Setting $H_0 : \rho_k = 0$ and $H_1 : \rho_k \neq 0$, we can define the following test statistic to evaluate the statistical significance of the correlation at a specific lag k:

$$t \text{ ratio} = \frac{\widehat{\rho}_k}{\sqrt{\left(1 + 2\sum_{i=1}^{k-1} \rho_i^2\right)/T}}$$

Under the Gaussian assumption, the t ratio is distributed as a normal distribution.

In finite samples, $\widehat{\rho}_k$ is a biased estimator of ρ (see Fisher [1915]). The bias is on the order of $1/T$, which can be quite large when the sample size T is small. However, in most financial applications, T is large enough that we do not have to worry about the impact of the bias.

Aside from evaluating the single correlation, we can assess whether the entire time series shows some dependency structure. To this purpose, we can use the Portmanteau test statistic

$$Q^*(m) = T \sum_{k=1}^{m} \widehat{\rho}_k^2 \qquad (1.4)$$

which is distributed as χ_m^2 under $H_0 \colon \rho_1 = \cdots = \rho_m = 0$ against $H_1 \colon \rho_k \neq 0$ for some $k \in \{1, \ldots, m\}$.

Ljung and Box (1978) modify (1.4) to increase the power of the test in finite samples as

$$Q^*(m) = T\,(T+2) \sum_{k=1}^{m} \frac{\widehat{\rho}_k^2}{T-k}$$

$Q^*(m)$ is asymptotically distributed as χ_{m-q}^2, which is a chi-squared distribution with $m - q$ degrees of freedom, where m is the number of coefficients estimated in the model. Because the value of m can affect the decision to reject or not to reject H_0, we usually check alternative values of m.

We can obtain the autocorrelation in two main ways. The first option is via the `corrgram` command, which produces the entire table of AC and PAC, along with the Portmanteau test statistic and a skinny plot of both AC and PAC. Applying `corrgram` to the S&P 500 returns, we get the following:

```
. corrgram return
```

| | | | | | -1   0   1 | -1   0   1 |
LAG	AC	PAC	Q	Prob>Q	[Autocorrelation]	[Partial Autocor]
1	0.0285	0.0285	13.061	0.0003		
2	-0.0421	-0.0429	41.559	0.0000		
3	0.0010	0.0035	41.575	0.0000		
4	-0.0040	-0.0059	41.828	0.0000		
5	-0.0134	-0.0130	44.742	0.0000		
6	-0.0052	-0.0049	45.18	0.0000		
7	-0.0188	-0.0197	50.895	0.0000		
8	0.0102	0.0109	52.557	0.0000		
9	-0.0064	-0.0088	53.218	0.0000		
10	0.0124	0.0137	55.711	0.0000		
11	-0.0132	-0.0150	58.503	0.0000		
12	0.0270	0.0287	70.269	0.0000		
13	0.0021	-0.0008	70.342	0.0000		
14	-0.0027	-0.0006	70.459	0.0000		
15	-0.0110	-0.0104	72.399	0.0000		
16	0.0327	0.0330	89.659	0.0000		
17	-0.0043	-0.0059	89.954	0.0000		
18	-0.0241	-0.0217	99.338	0.0000		
19	0.0003	0.0026	99.34	0.0000		
20	0.0114	0.0085	101.45	0.0000		
21	-0.0220	-0.0211	109.28	0.0000		
22	-0.0016	-0.0009	109.32	0.0000		
23	-0.0010	-0.0012	109.34	0.0000		
24	0.0101	0.0084	111	0.0000		
25	-0.0098	-0.0106	112.56	0.0000		
26	-0.0210	-0.0209	119.7	0.0000		

27	0.0185	0.0205	125.22	0.0000
28	0.0028	-0.0025	125.35	0.0000
29	0.0235	0.0256	134.25	0.0000
30	0.0091	0.0077	135.6	0.0000
31	-0.0105	-0.0074	137.39	0.0000
32	0.0180	0.0167	142.62	0.0000
33	0.0158	0.0150	146.64	0.0000
34	-0.0397	-0.0364	172.08	0.0000
35	0.0014	0.0052	172.11	0.0000
36	0.0031	-0.0010	172.26	0.0000
37	-0.0044	-0.0028	172.57	0.0000
38	-0.0064	-0.0043	173.23	0.0000
39	0.0112	0.0090	175.24	0.0000
40	-0.0079	-0.0097	176.25	0.0000

The results reported above show that our time series does not present any autocorrelation, with no peak being statistically significant and with the Portmanteau test rejecting the null hypothesis that autocorrelation is different from 0. After using the `corrgram` command, the autocorrelation, the partial autocorrelation, and the Q statistics are stored in the vectors `r(AC)`, `r(PAC)`, and `r(Q)`, respectively.

The second option allows us to generate just the autocorrelogram plot by using the `ac` command or the partial autocorrelogram by using the `pac` command. For instance, using the `ac` command on the S&P returns produces figure 1.11:

```
. ac return
```

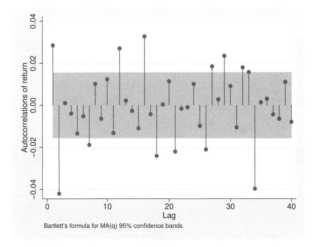

Figure 1.11. Autocorrelogram for S&P daily returns

Figure 1.11 reports the estimated autocorrelations denoted by filled circles connected by a straight line to the zero axis. The gray shadowed area identifies the α confidence level.

1.5.2   PACF

The partial autocorrelation for the process $\{r\}_t$ is given by the following:

$$\pi_k = \frac{E\left(r_t r_{t-k} | r_{t-1}, \dots, r_{t-k+1}\right)}{\text{var}\left(r_t\right)} \tag{1.5}$$

The main difference between the PACF and the ACF is that the former measures the linkage existing between time t and a generic lag $(t-k)$, regardless of observations occurring between $(t-k)$ and t. We can see that the numerator of (1.5) is the expected value of the covariance between time t and $(t-k)$, given $r_{t-1}, \dots, r_{t-k+1}$, which means pretending to be blinded about what happened between $(t-k)$ and t. Therefore, the PACF provides information about the additional contribution of each lagged value to the time structure of the time series. From an empirical point of view, the partial autocorrelation measures the linear relationship existing between the process at time t and $(t-k)$. We can estimate it by the coefficient π_k in the following regression:

$$r_t = \beta_0 + \beta_1 r_{t-1} + \dots + \beta_{t-k+1} r_{t-k+1} + \pi_k r_{t-k} + \epsilon_t$$

To illustrate this concept, we now run a simple regression of the S&P 500 returns on its five lags and compare the estimated coefficients with the first five lags of the PACF:

```
. regress return L.return L2.return L3.return L4.return L5.return
```

Source	SS	df	MS		Number of obs	=	16,097
					F(5, 16091)	=	9.25
Model	.004394413	5	.000878883		Prob > F	=	0.0000
Residual	1.52846952	16,091	.000094989		R-squared	=	0.0029
					Adj R-squared	=	0.0026
Total	1.53286394	16,096	.000095233		Root MSE	=	.00975

return	Coef.	Std. Err.	t	P>|t|	[95% Conf. Interval]	
return						
L1.	.0297561	.0078826	3.77	0.000	.0143052	.045207
L2.	-.0432508	.0078859	-5.48	0.000	-.0587081	-.0277935
L3.	.003074	.0078932	0.39	0.697	-.0123976	.0185456
L4.	-.0055074	.0078859	-0.70	0.485	-.0209647	.0099499
L5.	-.0130018	.0078823	-1.65	0.099	-.0284519	.0024483
_cons	.0002996	.000077	3.89	0.000	.0001487	.0004505

```
. corrgram return, lags(5)
                                                -1     0     1 -1      0        1
    LAG      AC       PAC        Q     Prob>Q  [Autocorrelation]  [Partial Autocor]

     1     0.0285    0.0285   13.061   0.0003
     2    -0.0421   -0.0429   41.559   0.0000
     3     0.0010    0.0035   41.575   0.0000
     4    -0.0040   -0.0059   41.828   0.0000
     5    -0.0134   -0.0130   44.742   0.0000
```

The two sets of coefficients are very close to each other, allowing us to conclude that the partial autocorrelation between time t and $(t-k)$ is identical to the marginal contribution of the $(t-k)$th lagged return in explaining returns at time t.

To obtain the PAC, we use the `pac` command on the S&P 500 returns:

```
. pac return
```

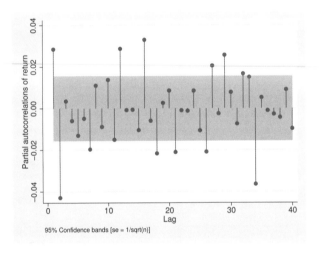

Figure 1.12. Partial autocorrelogram for S&P daily returns

The PAC shows that the time series of returns presents some significant peaks. For instance, the 1st, 7th, 14th, and 17th peaks pass the gray shadowed area, which identifies the area of no significant peaks.

1.6   Heteroskedasticity

When analyzing financial returns, it is quite unusual to find some time dependence. However, the picture changes when moving to squared returns.

Because we may assume the mean of returns to be very close to 0 without any loss of generality, the unconditional variance of returns can be computed as the average of the sum of squared returns,

$$\text{var}(r) = \frac{1}{T}\sum_{t=1}^{T}(r_t - \mu)^2 = \frac{1}{T}\sum_{t=1}^{T}r_t^2$$

where μ is the long-run mean of returns.

With the purpose of illustrating the behavior of squared returns, we now compute squared returns for the S&P 500 time series and draw its chart. We remove the Black Monday observation to better appreciate the time pattern:

```
. generate returns2 = return^2
(1 missing value generated)

. drop if date == td(19Oct1987)
(1 observation deleted)

. tsset date
        time variable:  date, 1/3/1950 to 12/31/2013, but with gaps
                delta:  1 day

. tsline returns2
```

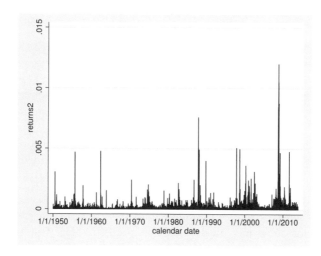

Figure 1.13. Squared returns for S&P 500 daily returns

We compute squared returns by using the **generate** command and setting the new **returns2** variable equal to the squared returns.

Observing the pattern of squared returns, we can easily identify periods of low volatility followed by periods of high volatility. For instance, we could recognize two periods of high volatility in recent history corresponding to the dot-com bubble in the early 2000s and the 2007–2012 financial and European Sovereign debt crisis. The two periods are separated by some years known as the Great Moderation, characterized by very low volatility. Periods of high volatility followed by periods of low volatility and vice versa is a common pattern in returns time series. Modeling this empirical evidence is the objective of an interesting stream of research, set up from the pioneer work on ARCH and GARCH models (see Engle [1982] and Bollerslev [1986]).

With the purpose of exploring conditional heteroskedasticity, we start by analyzing the autocorrelogram and partial autocorrelogram of squared returns.

```
use http://www.stata-press.com/data/feus/spdaily, clear
generate returns2 = return^2
tsset newdate
ac returns2, name(ac)
pac returns2, name(pac)
```

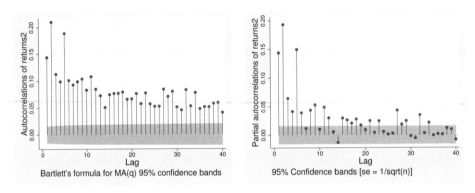

Figure 1.14. Autocorrelogram and partial autocorrelogram for squared S&P 500 daily returns

The autocorrelogram shows a high persistence of squared returns, a feature that could be nicely captured by an appropriate statistical model. We will devote chapter 3 to GARCH models, which were expressly introduced with that purpose.

A formal test for the presence of an ARCH effect is available after the **regress** command. The proposed test is a Lagrange multiplier test for the null hypothesis that the residuals do not show any autocorrelation pattern. The test is implemented as follows. In a first step, we obtain residuals; for instance, we can regress returns r_t on a constant μ so that we can obtain residuals as $\widehat{\varepsilon}_t = r_t - \widehat{\mu}$. In a second step, we regress squared residuals ε_t^2 on its p lags, fitting the following auxiliary regression:

$$\widehat{\varepsilon}_t^2 = \alpha_0 + \alpha_1 \widehat{\varepsilon}_{t-1}^2 + \cdots + \alpha_p \widehat{\varepsilon}_{t-p}^2 + u_t \tag{1.6}$$

On the basis of (1.6), we can introduce two test statistics,

1. $T R^2 \to \chi_p^2$

2. $F = \dfrac{R^2/p}{(1 - R^2)/(T - p)} \to F(p, T - p)$

where R^2 is the determination coefficient of the auxiliary regression in (1.6), T is the length of the residual time series, and p is the number of lags introduced in (1.6). The null hypothesis for both tests is that residuals do not show any heteroskedasticity feature: $H_0 : \alpha_1 = \alpha_2 = \cdots = \alpha_p = 0$.

We conduct the test for the presence of an ARCH effect on our S&P 500 returns as follows:

```
. regress return
```

Source	SS	df	MS			
				Number of obs	=	16,102
				F(0, 16101)	=	0.00
Model	0	0	.	Prob > F	=	.
Residual	1.53305434	16,101	.000095215	R-squared	=	0.0000
				Adj R-squared	=	0.0000
Total	1.53305434	16,101	.000095215	Root MSE	=	.00976

return	Coef.	Std. Err.	t	P>|t|	[95% Conf. Interval]	
_cons	.0002925	.0000769	3.80	0.000	.0001417	.0004432

```
. estat archlm, lags(1 5 10)
LM test for autoregressive conditional heteroskedasticity (ARCH)
```

lags(p)	chi2	df	Prob > chi2
1	332.085	1	0.0000
5	1347.482	5	0.0000
10	1441.739	10	0.0000

```
          H0: no ARCH effects      vs.  H1: ARCH(p) disturbance
```

In the first command, we run a simple regression with no covariates, just the constant. In the second command, we conduct the test by using `estat archlm`. We add into the auxiliary regression in (1.6) the 1st, 5th, and 10th lags. We can specify the lags in several ways; for instance, to test for lags 1 to k, we can type (1 2 3 ... k), (1/k), (1 2 to k), or (1 2:k).

The LM ARCH test strongly rejects the null hypothesis of no ARCH effect; all the lags added into the auxiliary regression in (1.6) are statistically significant.

We now want to investigate the relationship between resampling frequency and persistence. We therefore proceed by evaluating the autocorrelogram and the partial autocorrelogram of the S&P 500 monthly squared returns.

```
use http://www.stata-press.com/data/feus/spmonthly, clear
generate returns2 = return^2
ac returns2, name(acmonthly)
pac returns2, name(acmonthly)
```

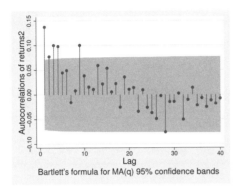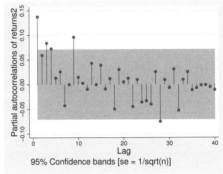

Figure 1.15. Autocorrelogram and partial autocorrelogram for squared S&P 500 monthly returns

The autocorrelogram and partial autocorrelogram show few significant correlation coefficients, especially if compared with figure 1.14, which reports the same charts but on daily data. We can therefore conclude that a positive relationship exists between persistence and resampling frequency: the finer the resampling frequency, the higher the persistence.

1.7   Linear time series

We can write a linear time series r_t as

$$r_t = \mu + \sum_{i=0}^{\infty} \psi_i \varepsilon_{t-i}$$

where μ is the long-run mean and ε_t is a Gaussian white-noise process $\varepsilon_t \sim$ i.i.d. $N(0, \sigma^2)$.

From an empirical point of view, the white-noise process ε_t represents the shocks or innovations arriving at time t and making returns r_t to deviate from their long-run mean.

Under the assumption that ε_t follows a white-noise process, we can obtain the mean and the variance of the process as

$$E\left(r_t\right) = E\left(\mu + \sum_{i=0}^{\infty} \psi_i \varepsilon_{t-i}\right) = \mu + E\left(\sum_{i=0}^{\infty} \psi_i \varepsilon_{t-i}\right) = \mu$$

$$\text{Var}\left(r_t\right) = \text{Var}\left(\mu + \sum_{i=0}^{\infty} \psi_i \varepsilon_{t-i}\right) = \sigma^2 \sum_{i=0}^{\infty} \psi_i$$

Linear models rely on assumptions that are too restrictive; they only allow for a linear dependence, that is, correlation.

We can instead represent nonlinear models as

$$r_t = f\left(a_t, a_{t-1}, a_{t-2}, \dots\right)$$

where $f(\cdot)$ is some nonlinear function. In particular, for modeling financial time series, we usually consider the following representation:

$$r_t = g\left(a_{t-1}, a_{t-2}, \dots\right) + a_t\, h\left(a_{t-1}, a_{t-2}, \dots\right)$$

where $g(\cdot)$ is the mean function and $h(\cdot)$ is the variance function.

1.8   Model selection

In general, when modeling financial time series, we undertake the following three steps:

1. model selection
2. model estimation
3. model testing

When selecting a model, we have to identify a family of models that satisfies the statistical properties of the financial time series we are modeling. For instance, if our data show the autocorrelation property, we have to select a class of models that account for this feature. After specifying the family of models, the next step is to fit the model on historical data. The final step, model testing, involves two parts: First, we need to evaluate the goodness of fit of our model by checking whether the model is correctly specified and whether it is able to explain the variability of our data. Second, we have to check the statistical significance of the estimated parameters; our model should at the same time be parsimonious and avoid overfitting. Overfitting occurs when a model is too heavily parameterized, compromising a model's efficiency and its applicability to out-of-sample data.

When there exist alternative models to describe the data-generating process, we need some rules to choose between them. A good model should satisfy two criteria: it should fit the data well and at the same time be parsimonious.

To this aim, the likelihood-ratio (LR) test compares two nested models, the restricted and the unrestricted models, with the unrestricted model having more parameters to be estimated than the restricted one. For instance, a regression with three covariates may act as the unrestricted version of a regression with just two of those three covariates, where the imposed restriction is that the third coefficient is set equal to 0. Denoting by Ψ the set of parameters for the restricted model and by $\ln L(\Psi)$ the corresponding log likelihood, and with Ω and $\ln L(\Omega)$ the same quantities for the unrestricted model, the LR test takes the following form:

$$\text{LR} = -2\left\{\ln L(\Psi) - \ln L(\Omega)\right\} \sim \chi^2_m$$

where χ_m^2 is the chi-squared distribution, with m degrees of freedom corresponding to the m restrictions that the restricted model imposes on the unrestricted one. The null hypothesis is that the two models are equivalent in terms of goodness of fit; when we cannot reject the null hypothesis, the restricted model should be preferred over the unrestricted one because it involves fewer parameters.

Keep in mind when comparing two or more models that we can use the LR test only when two models are nested. When two models are not nested, we have to adopt information criteria that take into account the goodness of fit, evaluated in terms of log likelihood, with the number of parameters used in the model. The two most common information criteria are the Akaike (AIC) and the Schwarz or Bayesian (BIC):

$$\mathrm{AIC} = -2\frac{\ln L}{T} + \frac{2}{T}k$$

$$\mathrm{BIC} = -2\frac{\ln L}{T} + \frac{\ln(T)}{T}k$$

where $\ln L$ is the log likelihood of the estimated model, k is the number of parameters, and T is the length of the time series. Between the two models, we should select the one that is characterized simultaneously by the higher log likelihood and the fewer parameters; therefore, the best model is always the one with the lowest information criteria, regardless of the criteria from which we are adopting.

The AIC tends to select the overparameterized model in comparison with the BIC, especially when the length of the time series T is short. In fact, in the AIC the penalty for each added parameter is equal to 2, while in the BIC the penalty is $\ln(T)$.

The underlying reason for that to happen is that the two information criteria are based on two different model selection approaches. The AIC is designed to identify the best approximating model to the unknown data-generating process, while the BIC is aimed at identifying the true model. Moreover, Bozdogan (1987) shows that the AIC lacks some properties of asymptotic consistency, while the BIC, being derived within a Bayesian framework, satisfies the asymptotic consistency property. This feature is reflected in the BIC applying a larger penalty than the AIC, which finally translates in the selection of less-parameterized models by the BIC when compared with the AIC.

The AIC and the BIC are useful and simple model selection criteria because they require us to fit the alternative models and choose the one with the smallest information criteria.

1.A How to import data

Stata allows users to import data from several alternative sources, including Excel; space-, tab-, or comma-separated sources; and SAS datasets. In this appendix, we describe how to import data in Excel format. We must first choose whether we want to proceed by using the user interface or by using an ad hoc command. We start here by describing the easiest way, which is to use the guided procedure that we can access by selecting **Excel spreadsheet (*.xls;*.xlsx)** from the **File ▷ Import** menu, as seen in figure 1.16.

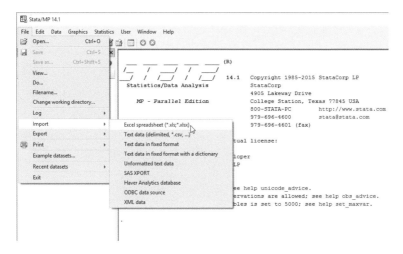

Figure 1.16. Screenshot of the **File** menu in Stata

The resulting window (see figure 1.17) asks us to indicate in **Excel file** where the Excel file is located. Then, Stata automatically proposes in the box labeled **Worksheet** the sheet and range with the data; we can edit the box **Cell range** to the appropriate range. If our data are labeled with a name in the first row, we may want to use those names as variable names; in that case, we should check the box for **Import first row as variable names**. The box labeled **Variable case** allows us to choose to preserve the name as it is (by selecting *Preserve*), to use capital letters (by selecting *Upper*), or to use lowercase letters (by selecting *Lower*). Finally, by checking the box for **Import all data as strings**, we can decide to set all the data as strings, although some of them can be numeric.

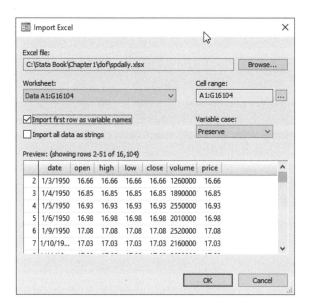

Figure 1.17. The Import Excel dialog box

Once we click on **OK**, Stata proposes the command to use to import our data:

```
import excel "http://www.stata-press.com/data/feus/spdaily.xlsx",   ///
        sheet("Data") firstrow clear
```

The `import excel` command is followed by the path where the Excel file is located. This is followed by the name of the worksheet containing the data and by the `firstrow` option, indicating that variable names are located in the first row of the range. The last option, `clear`, asks Stata to replace data in memory.

The `import excel` command has several additional interesting features. For instance, we could want to check our data before importing them. In that case, we can proceed as follows:

```
. import excel "http://www.stata-press.com/data/feus/spdaily.xlsx", describe
```

Sheet	Range
Data	A1:G16104

Moreover, we may want to import only some variable names, for instance, just `date` and `prices`:

```
import excel date prices using "http://www.stata-press.com/data/feus/spdaily.xlsx"
```

We now show how to generate new variables and set variable labels. First, we want to create a new date variable based on a calendar date, where the nonbusiness days are removed.

```
bcal create spdaily, from(date) generate(newdate) excludemissing(date) replace
tsset newdate
```

We add three other variables: `logprice`, corresponding to the logarithmic prices; `return`, computed as the difference in logarithmic prices; and `blackmonday`, a dummy variable taking a value of 1 to indicate the Black Monday:

```
generate logprice = ln(price)
generate return = d.logprice
generate blackmonday = 0
replace blackmonday = 1 if date==td(19Oct1987)
```

Finally, we want to set some labels to the variables:

```
label variable date "calendar date"
label variable date "business date"
label variable open "open price"
label variable high "high price"
label variable low "low price"
label variable close "close price"
label variable volume "traded volume"
label variable price "adj close"
label variable logprice "logarithmic price"
label variable return "logarithmic return"
```

2 ARMA models

According to the efficient market hypothesis, returns should not be predictable because all the available information is already discounted in the market price. Thus, traders should not be able to systematically beat the market because their forecasts should generally be in line with the market ones.

From a mathematical point of view, stock prices are martingales and satisfy the Markov property (as defined in chapter 1). This suggests that knowledge of the past does not contain any useful information for making predictions about future expected value.

However, some assets can show some degree of autocorrelation—in particular, the most illiquid ones—as a consequence of a slower price adjustment to news releases. In addition, some evidence shows that stock indexes are more characterized by the autocorrelation property than single stocks are (see, for instance, Chan [1993]).

2.1 Autoregressive (AR) processes

2.1.1 AR(1)

AR models are motivated by the high persistence that characterizes economic variables in general, although financial time series to a lesser extent. We can represent an AR process with one term, AR(1), as

$$r_t = \delta + \phi r_{t-1} + \varepsilon_t \tag{2.1}$$

where δ is the constant (accounting for the long-run mean), ϕ is the AR coefficient, and ε_t is an i.i.d. process. The value of ϕ identifies alternative processes. Specifically, when $\phi = 1$, the process in (2.1) is a random walk; when $|\phi| > 1$, the process gets explosive; and finally, when $|\phi| < 1$ (the case we are interested in), the process is stationary. We can rewrite (2.1) using the lag operator $L(\cdot)$. The lag operator $L(\cdot)$ allows us to shift the observation of a random variable, y, observed at time t at the previous lags $(t-1)$, $(t-2)$, ... :

$$L(y_t) = y_{t-1}$$
$$L^2(y_t) = y_{t-2}$$
$$\dots$$
$$L^p(y_t) = y_{t-p}$$

By means of the lag operator $L(\cdot)$, we can write the AR(1) process as

$$(1 - \phi L)\, r_t = \delta + \varepsilon_t$$

We can represent an AR(1) process as an infinite-order moving-average process, MA(∞). To obtain that representation, we proceed by substitution. At step $(t-1)$, we have

$$r_{t-1} = \delta + \phi r_{t-2} + \varepsilon_{t-1}$$

and by substituting it in (2.1), we get

$$r_t = \delta + \phi\left(\delta + \phi r_{t-2} + \varepsilon_{t-1}\right) + \varepsilon_t$$
$$r_t = \delta + \phi\delta + \phi^2 r_{t-2} + \phi\varepsilon_{t-1} + \varepsilon_t$$
$$r_t = \delta(1 + \phi) + \phi^2 r_{t-2} + \phi\varepsilon_{t-1} + \varepsilon_t$$

At step $(t-2)$, we have

$$r_{t-2} = \delta + \phi r_{t-3} + \varepsilon_{t-2}$$

which we can substitute in the previous step to obtain

$$r_t = \delta(1 + \phi + \phi^2) + \phi^3 r_{t-3} + \phi^2\varepsilon_{t-2} + \phi\varepsilon_{t-1} + \varepsilon_t$$

$$r_t = \delta \sum_{\tau=0}^{2} \phi^\tau + \phi^3 r_{t-3} + \sum_{\tau=0}^{2} \phi^\tau \varepsilon_{t-\tau}$$

Thus, after several substitutions, we get the following AR(1) process representation:

$$r_t = \delta \sum_{\tau=0}^{i-1} \phi^\tau + \phi^i r_{t-i} + \sum_{\tau=0}^{i-1} \phi^\tau \varepsilon_{t-\tau}$$

where $\delta \sum_{\tau=0}^{i-1} \phi^\tau$ is the long-run mean.

For $i \to \infty$ and under the stationarity condition $|\phi| < 1$, we get the infinite-order MA representation MA(∞) of an AR(1) process:

$$r_t = \lim_{i\to\infty} \delta \sum_{\tau=0}^{i-1} \phi^\tau + \phi^i r_{t-i} + \sum_{\tau=0}^{i-1} \phi^\tau \varepsilon_{t-\tau} = \frac{\delta}{1-\phi} + \sum_{\tau=0}^{\infty} \phi^\tau \varepsilon_{t-\tau} \qquad (2.2)$$

Therefore, we can represent an AR(1) model as a linear combination of i.i.d. variables, with exponential weights ϕ^τ. From the MA(∞) representation in (2.2), we can see that the AR(1) process has finite memory, because the current value of the process r_t depends on past values, though the weights of past values decrease exponentially. The MA(∞) representation of a stationary AR(1) process relies on the Wold theorem, which states that all the covariance-stationary processes can be decomposed as a sum of two time series, one deterministic and one stochastic. Formally, this means

$$r_t = \mu + \sum_{\tau=0}^{\infty} \psi_\tau \varepsilon_{t-\tau} \qquad (2.3)$$

where μ_t is the deterministic part and $\sum_{\tau=0}^{\infty} \psi_\tau \varepsilon_{t-\tau}$ is the stochastic part, and where ε_t is the white-noise process and ψ is the MA weights such that $\psi_0 = 1$ and $\sum_{\tau=0}^{\infty} |\psi_\tau|^2 < \infty$. The Wold theorem is useful because it allows the approximation of the dynamic evolution of a variable r_t by a linear model. If the innovations ε_t are independent, then the linear model is the only possible representation relating the observed value of r_t to its past evolution.

Having presented the Wold theorem, it is straightforward to introduce the impulse–response function. The impulse–response function allows us to investigate the response of one variable to an impulse in another variable in a system that involves multiple variables as well.

We can see weights as an impulse response:

$$\frac{\partial r_{t+s}}{\partial \varepsilon_t} = \phi_t \qquad t = 1, \dots, T$$

We can obtain the impulse–response function by simply plotting the weights ϕ_t against the time. In such a way, we can assess the response of the dependent variable to a unit shock.

Under the assumption that the process r_t in (2.3) is stationary, for example, when ε_t is i.i.d. and exploiting the MA(∞) representation of the AR(1) process, we can show that by imposing the long-run mean $\delta = 0$, the expected value of an AR(1) process is 0:

$$E\left(r_t\right) = E\left(\sum_{\tau=0}^{\infty} \phi^\tau \varepsilon_{t-\tau}\right) = 0$$

In addition, the variance is equal to a constant γ_0:

$$\begin{aligned}
\text{var}\left(r_t\right) &= \text{var}\left(\sum_{\tau=0}^{\infty} \phi^\tau \varepsilon_{t-\tau}\right) \\
\text{var}\left(r_t\right) &= \sum_{\tau=0}^{\infty} \left(\phi^\tau\right)^2 \text{var}\left(\varepsilon_{t-\tau}\right) \\
\text{var}\left(r_t\right) &= \sum_{\tau=0}^{\infty} \left(\phi^\tau\right)^2 \sigma^2 \\
\text{var}\left(r_t\right) &= \frac{\sigma^2}{1-\phi^2} = \gamma_0
\end{aligned}$$

The condition for the existence of the second noncentral moment is that $\phi^2 < 1$. We note that, together with the assumption of weak stationarity, the mean and the variance of an AR(1) process do not depend on t.

By including the constant δ, we can show that the expected value of the AR(1) model would get

$$\begin{aligned}
E\left(r_t\right) &= E\left(\delta + \phi r_{t-1} + \varepsilon_t\right) \\
\mu &= \delta + \phi\mu \\
\mu &= \frac{\delta}{(1-\phi)}
\end{aligned}$$

We now derive the theoretical AC for an AR(1) process:

$$\gamma_1 = \operatorname{cov}\left(r_t, r_{t-1}\right)$$
$$\gamma_1 = E\left\{\left(\phi r_{t-1} + \varepsilon_t\right)\left(r_{t-1}\right)\right\}$$
$$\gamma_1 = E\left(\phi r_{t-1} r_{t-1} + \varepsilon_t r_{t-1}\right)$$
$$\gamma_1 = E\left(\phi r_{t-1} r_{t-1}\right) + E\left(\varepsilon_t r_{t-1}\right)$$
$$\gamma_1 = \phi \operatorname{var}\left(r_{t-1}\right)$$
$$\gamma_1 = \phi \gamma_0$$

$$\gamma_2 = \operatorname{cov}\left(r_t, r_{t-2}\right)$$
$$\gamma_2 = E\left\{\left(\phi^2 r_{t-2} + \phi \varepsilon_{t-1} + \varepsilon_t\right)\left(r_{t-2}\right)\right\}$$
$$\gamma_2 = E\left(\phi^2 r_{t-2} r_{t-2} + \phi \varepsilon_{t-1} r_{t-2} + \varepsilon_t r_{t-2}\right)$$
$$\gamma_2 = E\left(\phi^2 r_{t-2} r_{t-2}\right) + E\left(\phi \varepsilon_{t-1} r_{t-2}\right) + E\left(\varepsilon_t r_{t-2}\right)$$
$$\gamma_2 = \phi^2 \operatorname{var}\left(r_{t-2}\right)$$
$$\gamma_2 = \phi^2 \gamma_0$$

Proceeding in a similar way, we could show that at the generic kth lag, the covariance γ_k is equal to the following:

$$\gamma_k = \phi^k \gamma_0$$

Having in hand the formula for the autocovariance, we can show that the AC for an AR(1) process is

$$\rho_0 = \tfrac{\gamma_0}{\gamma_0} = 1$$

$$\rho_1 = \tfrac{\gamma_1}{\gamma_0} = \phi$$

$$\rho_2 = \tfrac{\gamma_2}{\gamma_0} = \phi^2$$

$$\cdots$$

$$\rho_k = \tfrac{\gamma_k}{\gamma_0} = \phi^k$$

The above expression of the AC can be useful when we need to identify the form of an empirical process. According to the above representation, the theoretical AC of an AR(1) process decreases exponentially, changing sign if $-1 < \phi < 0$, while always being positive if $\phi > 0$.

To explore the properties of an AR(1) model, we proceed by simulating some AR(1) processes with alternative autoregressive coefficients and investigating their ACF and PACF. The first step is to simulate an AR(1) task, which can be achieved by the simple program reported below:

```
set obs 1500
* fix the seed
set seed 1
```

```
generate time = _n
tsset time

* define parameters for simulation
scalar constant = 0.01
scalar AR = 0.9

* simulate the process
generate mynormal = rnormal()

* AR(0.9)
generate y = .
replace y = mynormal in 1
replace y = constant + AR*L.y + mynormal if time>1
label variable y "AR(0.9)"
drop if _n < 500
drop if _n < 500
```

In the first part, we set the parameters for the simulations, namely, the constant `constant` and the AR(1) parameter `AR`, with the length of the simulation period (`obs 1500`). We even create a time variable by typing `time=_n`, which generates a count variable from 1 to the length of the simulation, previously fixed and equal to `_n`, which is an internal Stata variable. At this stage, we indicate that the data we are using is a time series by using the `tsset` command followed by the name of the time variable, which in our case is `time`. In addition, we set the seed (the starting point for any random sequence) to ensure to get the same sequence of random numbers every time the simulation is run.

The second part of the code refers to the core of the simulation. We extract random values ε_t from a standard univariate normal distribution by using the `rnormal()` function and store the values in the new `mynormal` variable. To create new variables, we use the `generate` command. Having generated the random variable `mynormal` and set the `time`, we can simulate the AR(1) process. We first initialize the y time series equal to an empty variable to afterward simulate the proper AR(1) model. The time-series operator `L.y` indicates that we are using the lag operator. To ensure that the simulated time series will have the desired properties, we generate a larger number of observations than we need (a common practice known as burn in the draws). Generating a larger number of time points than we actually need helps in making draws closer to the stationary distribution and less dependent on the starting point as the effect of the initial condition dies out for stationary processes.

When deciding the number of points to drop, we should take care because discarding a too short burn-in period will leave some problematic observations, while discarding a too large burn-in period will throw away representative observations that could be helpful in improving accuracy for the parameter estimates. In our case, we have generated 1,500 observations, but we drop the first 499 observations: `drop if _n < 500`.

We now show in figure 2.1 plots of the AR(1) process for alternative values of the AR parameter.

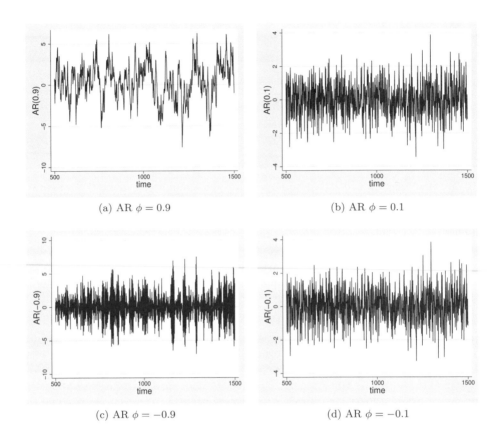

(a) AR $\phi = 0.9$ (b) AR $\phi = 0.1$

(c) AR $\phi = -0.9$ (d) AR $\phi = -0.1$

Figure 2.1. Simulations of AR(1) process with alternative AR parameters

From figure 2.1, we focus our attention on the AR(0.9) process. We see that the runs above or below the mean are longer than we would observe in an i.i.d. process and that the process is smoother than an i.i.d. process. In contrast, in the AR(-0.9) process, we note that the runs above or below the mean are shorter than we would observe in an i.i.d. process and that the process is more jagged than an i.i.d. process.

Having simulated alternative AR(1) processes, it is interesting now to compare ACs and PACs (see figure 2.2) by using the `corrgram`, `ac`, and `pac` commands as described in chapter 1. We start with the AR(1) with an autoregressive parameter ϕ equal to 0.9.

```
. corrgram y, lags(20)
```

| | | | | | -1   0   1 -1   0   1 |
LAG	AC	PAC	Q	Prob>Q	[Autocorrelation] [Partial Autocor]
1	0.8939	0.8943	802.32	0.0000	
2	0.7939	-0.0272	1435.7	0.0000	
3	0.6970	-0.0409	1924.5	0.0000	
4	0.6127	0.0068	2302.6	0.0000	
5	0.5364	-0.0111	2592.6	0.0000	
6	0.4767	0.0344	2821.9	0.0000	
7	0.4291	0.0238	3007.8	0.0000	
8	0.3925	0.0284	3163.6	0.0000	
9	0.3571	-0.0115	3292.7	0.0000	
10	0.3319	0.0374	3404.3	0.0000	
11	0.3020	-0.0313	3496.8	0.0000	
12	0.2776	0.0161	3575.1	0.0000	
13	0.2685	0.0722	3648.3	0.0000	
14	0.2626	0.0182	3718.5	0.0000	
15	0.2508	-0.0236	3782.5	0.0000	
16	0.2405	0.0102	3841.5	0.0000	
17	0.2318	0.0143	3896.3	0.0000	
18	0.2135	-0.0441	3942.9	0.0000	
19	0.1818	-0.0663	3976.7	0.0000	
20	0.1506	-0.0122	3999.9	0.0000	

```
. ac y, lags(20) name(ACAR09)

. pac y, lags(20) name(PACAR09)
```

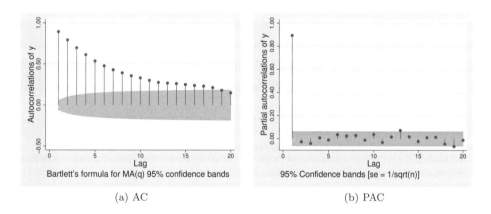

(a) AC   (b) PAC

Figure 2.2. ACF and PACF for the AR(1) process with AR parameter equal to 0.9

The AC shows a decaying pattern. In the PAC, only the first peak is positive and statistically significant, lying outside the gray shaded area. As stated before, this is the typical pattern that characterizes an AR(1) process. In addition, the Portmanteau test statistic, identified by the label Q in the output above, shows that we reject the null hypothesis that the autocorrelations are equal to 0. One way to check that our model is correctly specified is to evaluate whether we have averaged out the autocorrelation initially characterizing the time series. Thus, we focus our attention on residuals because now they should follow a white-noise process. In our case, because we are working on simulated data, we know that the true model is an AR(1), and so we expect residuals to follow a white-noise process. We first fit an AR(1) model on the simulated data, and afterward, we store residuals in a new **res** variable.

```
. arima y, ar(1)
  (output omitted)
. predict res, residuals
```

Finally, we analyze the correlogram for residuals:

```
. corrgram res, lags(20)
```

| | | | | | -1   0   1 -1   0   1 |
LAG	AC	PAC	Q	Prob>Q	[Autocorrelation] [Partial Autocor]
1	0.0252	0.0252	.6365	0.4250	
2	0.0323	0.0317	1.6818	0.4313	
3	-0.0170	-0.0187	1.9728	0.5781	
4	0.0005	0.0003	1.9731	0.7407	
5	-0.0426	-0.0416	3.804	0.5780	
6	-0.0258	-0.0242	4.4736	0.6129	
7	-0.0267	-0.0229	5.196	0.6361	
8	0.0167	0.0182	5.4777	0.7055	
9	-0.0280	-0.0287	6.2731	0.7123	
10	0.0413	0.0398	8.0053	0.6283	
11	-0.0084	-0.0102	8.0777	0.7063	
12	-0.0506	-0.0572	10.679	0.5566	
13	-0.0001	0.0047	10.679	0.6377	
14	0.0410	0.0436	12.393	0.5748	
15	0.0073	0.0062	12.448	0.6449	
16	0.0068	0.0030	12.494	0.7093	
17	0.0558	0.0563	15.676	0.5469	
18	0.0711	0.0650	20.846	0.2872	
19	0.0063	0.0027	20.887	0.3431	
20	-0.0052	-0.0051	20.915	0.4021	

As expected, the residuals do not show any evidence of autocorrelation, with both the AC and the PAC presenting no pick, and the Portmanteau test statistics leading us to not reject the null hypothesis of absence of autocorrelation. We can conclude that the AR(1) is the correct model for our time series because it is able to capture all the features of the time series.

We now report a similar analysis for the case of an AR(1) model with a negative autoregressive parameter, $\phi = -0.9$ (see figure 2.3).

```
. corrgram y2, lags(20)
```

					-1 0 1 -1 0 1
LAG	AC	PAC	Q	Prob>Q	[Autocorrelation] [Partial Autocor]
1	-0.9097	-0.9115	830.85	0.0000	
2	0.8324	0.0347	1527.3	0.0000	
3	-0.7598	0.0101	2108.1	0.0000	
4	0.6918	-0.0102	2590	0.0000	
5	-0.6328	-0.0163	2993.7	0.0000	
6	0.5712	-0.0368	3322.9	0.0000	
7	-0.5218	-0.0359	3597.9	0.0000	
8	0.4732	-0.0181	3824.3	0.0000	
9	-0.4270	0.0095	4008.8	0.0000	
10	0.3810	-0.0317	4155.9	0.0000	
11	-0.3326	0.0440	4268.1	0.0000	
12	0.2861	-0.0264	4351.2	0.0000	
13	-0.2537	-0.0505	4416.6	0.0000	
14	0.2242	-0.0019	4467.7	0.0000	
15	-0.1895	0.0493	4504.3	0.0000	
16	0.1578	-0.0095	4529.7	0.0000	
17	-0.1270	0.0158	4546.1	0.0000	
18	0.1080	0.0443	4558	0.0000	
19	-0.0768	0.0774	4564.1	0.0000	
20	0.0483	-0.0185	4566.5	0.0000	

```
. ac y2, lags(20) name(ACARNeg09)

. pac y2, lags(20) name(PACARNeg09)
```

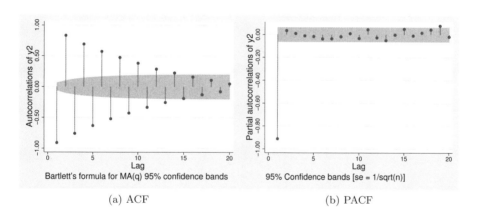

(a) ACF   (b) PACF

Figure 2.3. ACF and PACF for the AR(1) process with AR parameter equal to −0.9

The AC shows a decaying pattern, changing signs from lag to lag. In the PAC, only the first peak is negative and statistically significant. This is the typical pattern that the AC and PAC follow in the presence of an AR(1) model with a negative autoregressive coefficient.

As before, we now verify that the residuals follow a white-noise process.

```
. drop res
. arima y2, ar(1)
  (output omitted)
. predict res, residuals
. corrgram res, lags(20)
```

LAG	AC	PAC	Q	Prob>Q	-1 0 1 [Autocorrelation]	-1 0 1 [Partial Autocor]
1	0.0305	0.0305	.93134	0.3345		
2	0.0175	0.0166	1.2379	0.5385		
3	-0.0120	-0.0130	1.3819	0.7098		
4	-0.0137	-0.0133	1.571	0.8140		
5	-0.0397	-0.0386	3.1592	0.6755		
6	-0.0361	-0.0338	4.4744	0.6128		
7	-0.0237	-0.0208	5.0419	0.6549		
8	0.0077	0.0091	5.1017	0.7467		
9	-0.0261	-0.0282	5.7921	0.7605		
10	0.0345	0.0334	6.9965	0.7258		
11	-0.0070	-0.0111	7.0463	0.7953		
12	-0.0581	-0.0631	10.468	0.5750		
13	0.0019	0.0050	10.472	0.6550		
14	0.0364	0.0385	11.821	0.6207		
15	0.0077	0.0049	11.881	0.6880		
16	0.0031	-0.0001	11.891	0.7515		
17	0.0562	0.0553	15.109	0.5876		
18	0.0692	0.0633	19.999	0.3329		
19	0.0058	0.0024	20.034	0.3925		
20	-0.0055	-0.0033	20.065	0.4539		

Residuals from the AR(1) model follow a white-noise process, with both the AC and the PAC presenting no statistically significant autocorrelation for all 20 lags considered and the Portmanteau statistics leading us to not reject the null hypothesis of no autocorrelation. We can therefore conclude that the model is correctly specified.

2.1.2 AR(p)

An AR(p) process is a generalization of the AR(1) process to p lags:

$$r_t = \delta + \phi_1 r_{t-1} + \phi_2 r_{t-2} + \cdots + \phi_p r_{t-p} + \varepsilon_t \tag{2.4}$$

with ε_t being a white-noise process, assumed to follow a Gaussian distribution, with a mean of 0 and a constant variance σ^2.

Similarly to what we did for the AR(1) process, we now proceed by obtaining the expected value for an AR(p) model. The expected value of the AR(p) process with a nonzero constant δ is

$$E\left(r_t\right) = \frac{\delta}{1 - \phi_1 - \phi_2 - \cdots - \phi_p}$$

where the condition that $\phi_1 + \phi_2 + \cdots + \phi_p < 1$ must be satisfied to guarantee that the process is well defined. The polynomial equation associated with the model in (2.4), also known as the characteristics equation, is

$$1 - \phi_1 r - \phi_2 r^2 - \cdots - \phi_p r^p = 0 \tag{2.5}$$

The stationarity condition that an AR process with p terms must fulfill is that all the solutions of (2.5) are greater than 1 in modulus. The inverses of the solutions are called the characteristic roots of the model. We could define the stationarity condition also in terms of characteristic roots that, for symmetry, must be smaller than 1 in modulus.

Using the lag operator, we can rewrite the AR(p) process as

$$\left(1 - \phi_1 L - \phi_2 L^2 - \cdots - \phi_p L^p\right) r_t = \delta + \varepsilon_t$$

In addition to that, we can define a lag polynomial as

$$\phi\left(L\right) = \left(1 - \phi_1 L - \phi_2 L^2 - \cdots - \phi_p L^p\right)$$

and rewrite the AR(p) process as

$$\phi\left(L\right) r_t = \delta + \varepsilon_t$$

2.2 Moving-average (MA) processes

2.2.1 MA(1)

The idea underlying MA processes is that the level of a random variable at time t is affected not only by a shock occurring at time t but also at previous times, with the number of significant lags defining the order of the process. The assumption is that when a negative shock to the economy occurs—say, the default of a financial institution—we expect this negative news to affect financial markets not only at the time it takes place but also at times in the near future.

An MA process with one term, MA(1), is defined as

$$r_t = \mu + \theta \varepsilon_{t-1} + \varepsilon_t \tag{2.6}$$

where μ is the constant term and ε_t is a Gaussian white-noise process with mean 0 and constant variance σ^2.

Using the lag operator, we can rewrite the MA(1) process as

$$r_t = \mu + (1 + \theta L)\varepsilon_t$$

The expected value of an MA(1) process is equal to the constant term μ,

$$E\left(r_t\right) = E\left(\mu + \theta \varepsilon_{t-1} + \varepsilon_t\right) = \mu$$

where we have used the assumption that $E(\varepsilon_t) = 0$ and that ε_t and ε_{t-1} are independent. As for the variance, we can proceed as follows:

$$\text{var}(r_t) = \text{var}(\mu + \theta\varepsilon_{t-1} + \varepsilon_t)$$
$$\text{var}(r_t) = \theta^2\text{var}(\varepsilon_{t-1}) + \text{var}(\varepsilon_t)$$
$$\text{var}(r_t) = \theta^2\sigma^2 + \sigma^2$$
$$\text{var}(r_t) = \sigma^2(\theta^2 + 1)$$

In the derivation, we have exploited again the property of independence characterizing the ε_t process, implying $\text{cov}(\varepsilon_t, \varepsilon_{t-1}) = 0$. In addition, we have $\text{var}(\varepsilon_t) = \sigma^2$ for the assumption stated above about the distribution of ε_t.

Finally, the covariance between r_t and r_{t-1} of an MA(1) process is

$$\gamma_1 = \text{cov}(r_t, r_{t-1})$$
$$\gamma_1 = \text{cov}\{(\mu + \theta\varepsilon_{t-1} + \varepsilon_t), (\mu + \theta\varepsilon_{t-2} + \varepsilon_{t-1})\}$$
$$\gamma_1 = E\{(\mu + \theta\varepsilon_{t-1} + \varepsilon_t)(\mu + \theta\varepsilon_{t-2} + \varepsilon_{t-1})\} - E(\mu + \theta\varepsilon_{t-1} + \varepsilon_t)$$
$$\qquad E(\mu + \theta\varepsilon_{t-2} + \varepsilon_{t-1})$$
$$\gamma_1 = E\left(\theta^2\varepsilon_{t-1}\varepsilon_{t-2} + \theta\varepsilon_{t-1}^2 + \theta\varepsilon_t\varepsilon_{t-2} + \varepsilon_t\varepsilon_{t-1}\right)$$
$$\gamma_1 = \theta\sigma^2$$

We can show that for an MA(1) process, the covariance between r_t and r_{t-2} is equal to 0:

$$\gamma_2 = \text{cov}(r_t, r_{t-2})$$
$$\gamma_2 = \text{cov}\{(\mu + \theta\varepsilon_{t-1} + \varepsilon_t)(\mu + \theta\varepsilon_{t-3} + \varepsilon_{t-2})\}$$
$$\gamma_2 = E\{(\mu + \theta\varepsilon_{t-1} + \varepsilon_t)(\mu + \theta\varepsilon_{t-3} + \varepsilon_{t-2})\} - E(\mu + \theta\varepsilon_{t-1} + \varepsilon_t)$$
$$\qquad E(\mu + \theta\varepsilon_{t-3} + \varepsilon_{t-2})$$
$$\gamma_2 = E\left(\theta^2\varepsilon_{t-1}\varepsilon_{t-3} + \theta\varepsilon_{t-1}\varepsilon_{t-2} + \theta\varepsilon_t\varepsilon_{t-3} + \varepsilon_t\varepsilon_{t-2}\right)$$
$$\gamma_2 = 0$$

The process is stationary given that the expected value, the variance, and the covariance do not depend on t. This result holds true without imposing any condition on θ. The stationarity property is ensured by the model specification being a linear combination of white-noise processes.

Having computed the variance and the covariance for an MA(1) process, we are ready to calculate the theoretical AC as follows:

$$\rho_0 = \frac{\gamma_0}{\gamma_0} = 1$$

$$\rho_1 = \frac{\gamma_1}{\gamma_0} = \frac{\theta}{(\theta^2 + 1)}$$

$$\rho_2 = \frac{\gamma_2}{\gamma_0} = 0$$

$$\cdots$$

$$\rho_k = \frac{\gamma_k}{\gamma_0} = 0$$

It is worth noticing that the AC is a short memory process, with the coefficient getting equal to 0 at the second lag.

Similarly to what we did for the AR(1) process, we now simulate some examples of MA(1) processes with alternative values of θ. We report below the code used for simulation:

```
clear all
set obs 1500

* fix the seed
set seed 1

generate time = _n
tsset time

* define parameters for simulation
scalar constant = 0.01
scalar MA = 0.9

* simulate the process
generate mynormal = rnormal()

* MA(0.9)
generate y = .
replace y = mynormal in 1
replace y = constant + MA*L.mynormal + mynormal if time>1
label variable y "MA(0.9)"
drop if time < 500
```

The code is similar to that used for simulating an AR(1) process. In the first part of the code, we specify the parameters, while in the second part, we extract random numbers from a normal distribution. Finally, in the last part, we generate the MA(1) process, dropping the first 499 observations.

In figure 2.4, we report plots of the MA(1) process for alternative values of the θ coefficient.

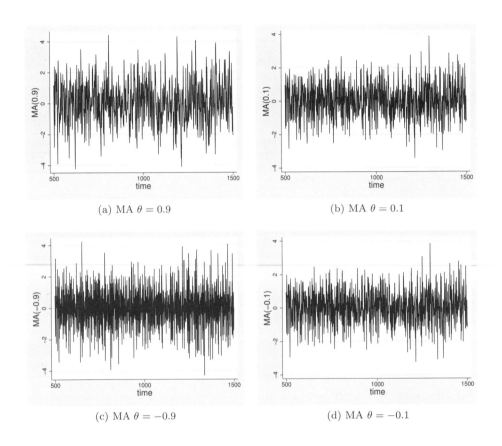

(a) MA $\theta = 0.9$ (b) MA $\theta = 0.1$

(c) MA $\theta = -0.9$ (d) MA $\theta = -0.1$

Figure 2.4. Simulations of MA(1) process with alternative MA parameters

By comparing figure 2.4 with figure 2.1, we can see that the MA(1) processes are much closer to white-noise processes than are the AR(1) processes. In fact, MA(1) processes are a simple function of innovations at the previous time point, so it is quite hard to identify a time pattern. Moreover, we can see that for MA with 0.9 and −0.9 coefficients, runs above and below the mean are longer than those for MA with −0.1 and 0.1 coefficients.

Having in hand the simulated MA processes, we can analyze the ACF and the PACF, taking as an example the case of $\theta = 0.9$ (see figure 2.5).

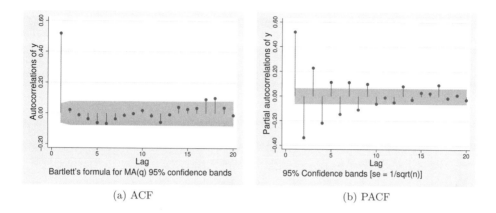

(a) ACF   (b) PACF

Figure 2.5. ACF and PACF for MA(1) process with MA parameter equal to 0.9

The AC has only one positive and statistically significant coefficient, while the PAC decreases exponentially changing sign. Figure 2.5 reports the typical shape of the AC and the PAC of an MA(1) process with a positive coefficient. We now verify that the residuals from the MA(1) model follow a white-noise process.

```
. arima y, ma(1)
  (output omitted)
. predict res, residuals
. corrgram res, lags(20)
```

| | | | | | -1   0   1 | -1   0   1 |
LAG	AC	PAC	Q	Prob>Q	[Autocorrelation]	[Partial Autocor]
1	0.0340	0.0341	1.1622	0.2810		
2	0.0112	0.0101	1.2883	0.5251		
3	-0.0115	-0.0124	1.4222	0.7003		
4	-0.0124	-0.0119	1.5773	0.8129		
5	-0.0373	-0.0362	2.9781	0.7034		
6	-0.0403	-0.0380	4.62	0.5934		
7	-0.0222	-0.0192	5.116	0.6458		
8	0.0073	0.0088	5.1703	0.7392		
9	-0.0242	-0.0265	5.7615	0.7635		
10	0.0353	0.0345	7.0232	0.7233		
11	-0.0051	-0.0097	7.0493	0.7951		
12	-0.0568	-0.0615	10.326	0.5874		
13	0.0003	0.0035	10.326	0.6671		
14	0.0384	0.0403	11.825	0.6204		
15	0.0084	0.0054	11.897	0.6868		
16	0.0032	0.0007	11.907	0.7503		
17	0.0592	0.0585	15.478	0.5611		
18	0.0708	0.0643	20.595	0.3003		
19	0.0057	0.0023	20.629	0.3577		
20	-0.0099	-0.0072	20.729	0.4133		

As expected, neither the AC nor the PAC show statistically significant peaks, indicating that the MA(1) model has successfully extracted the autocorrelation affecting the original time series.

We now report a similar analysis for the case of an MA(1) process with the negative parameter $\theta = -0.9$ (see figure 2.6).

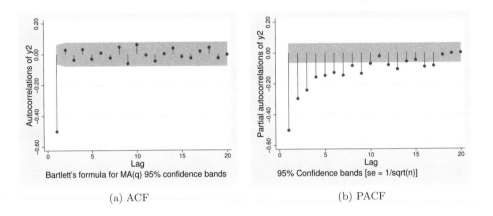

| (a) ACF | (b) PACF |

Figure 2.6. ACF and PACF for MA(1) process with MA parameter equal to -0.9

Similarly to the case of the MA(1) with $\theta = 0.9$, here in figure 2.6 the AC has only one significant but negative coefficient, while the PAC decreases exponentially. The patterns of the AC and the PAC together identify an MA(1) process with a negative coefficient. Again, residuals now follow a white-noise process.

```
. arima y2, ma(1)
  (output omitted )
. predict res2, residuals
. corrgram res2, lags(20)
```

| | | | | | -1   0   1 | -1   0   1 |
LAG	AC	PAC	Q	Prob>Q	[Autocorrelation]	[Partial Autocor]
1	0.0244	0.0245	.59897	0.4390		
2	0.0288	0.0283	1.4336	0.4883		
3	-0.0201	-0.0216	1.8402	0.6062		
4	0.0012	0.0013	1.8417	0.7649		
5	-0.0424	-0.0413	3.6519	0.6005		
6	-0.0279	-0.0266	4.4401	0.6173		
7	-0.0277	-0.0240	5.2139	0.6339		
8	0.0163	0.0176	5.481	0.7051		
9	-0.0290	-0.0298	6.3299	0.7065		
10	0.0409	0.0392	8.0234	0.6266		
11	-0.0084	-0.0102	8.0951	0.7048		
12	-0.0500	-0.0567	10.63	0.5609		
13	-0.0021	0.0024	10.634	0.6414		
14	0.0414	0.0436	12.377	0.5760		
15	0.0073	0.0060	12.432	0.6461		
16	0.0052	0.0016	12.459	0.7118		
17	0.0561	0.0568	15.666	0.5476		
18	0.0724	0.0663	21.023	0.2782		
19	0.0080	0.0047	21.089	0.3319		
20	-0.0070	-0.0065	21.138	0.3890		

2.2.2   MA(q)

We can easily generalize the MA(1) process to an MA(q) process including q lags:

$$r_t = \mu + \theta_1 \varepsilon_{t-1} + \theta_2 \varepsilon_{t-2} + \cdots + \theta_q \varepsilon_{t-q} + \varepsilon_t \tag{2.7}$$

where ε_t is the usual Gaussian white-noise process with a mean of 0 and a constant variance of σ^2. Using the lag operator $L(\cdot)$, we can rewrite the MA(q) process in (2.7) as

$$r_t = \mu + (1 + \theta_1 L + \theta_2 L^2 + \cdots + \theta_q L^q) \varepsilon_t$$

Or, by using the lag polynomial $\theta(L) = 1 + \theta_1 L + \theta_2 L^2 + \cdots + \theta_q L^q$, we can rewrite the process in (2.7) as

$$r_t = \mu + \theta(L) \varepsilon_t$$

The expected value of an MA(q) process is equal to the constant μ,

$$E(r_t) = E(\mu + \theta_1 \varepsilon_{t-1} + \theta_2 \varepsilon_{t-2} + \cdots + \theta_q \varepsilon_{t-q} + \varepsilon_t) = \mu$$

while the variance takes the following form:

$$
\begin{aligned}
\text{var}\,(r_t) &= \text{var}\,(\mu + \theta_1\varepsilon_{t-1} + \theta_2\varepsilon_{t-2} + \cdots + \theta_q\varepsilon_{t-q} + \varepsilon_t)\\
\text{var}\,(r_t) &= \text{var}\,(\mu) + \theta_1^2\text{var}\,(\varepsilon_{t-1}) + \theta_2^2\text{var}\,(\varepsilon_{t-2}) + \cdots + \theta_q^2\text{var}\,(\varepsilon_{t-q}) + \text{var}\,(\varepsilon_t)\\
\text{var}\,(r_t) &= \theta_1^2\sigma^2 + \theta_2^2\sigma^2 + \cdots + \theta_q^2\sigma^2 + \sigma^2\\
\text{var}\,(r_t) &= \sigma^2\left(\theta_1^2 + \theta_2^2 + \cdots + \theta_q^2 + 1\right)
\end{aligned}
$$

The variance of an MA(q) process is well defined if $\sum_{i=1}^{q}\theta_i^2 < \infty$.

We skip the computation of the autocovariance, because this would require some tedious calculus. Instead, we present the AC for the kth lag:

$$
\rho_k = \frac{\gamma_k}{\gamma_0} = \begin{cases} \frac{\theta}{1+\theta^2} & if\ k \le q \\ 0 & if\ k > q \end{cases} \tag{2.8}
$$

The function of the AC in (2.8) indicates that the memory of an MA(q) process lasts for q lags, while it is equal to 0 for lags greater than q. Therefore, as done previously, we can use this result to identify the best process for an empirical time series.

2.2.3 Invertibility

We can represent an MA process as an AR process provided that the invertibility condition is satisfied. To illustrate this point, we can start by rewriting the MA(1) process in (2.6) as

$$
\varepsilon_t = r_t + \theta\varepsilon_{t-1}
$$

Proceeding by substitutions:

$$
\begin{aligned}
\varepsilon_{t-1} &= r_{t-1} + \theta\varepsilon_{t-2} \rightarrow \varepsilon_t = r_t + \theta r_{t-1} + \theta^2\varepsilon_{t-2}\\
\varepsilon_{t-2} &= r_{t-2} + \theta\varepsilon_{t-3} \rightarrow \varepsilon_t = r_t + \theta r_{t-1} + \theta^2 r_{t-2} + \theta^3\varepsilon_{t-3}
\end{aligned}
$$

We can obtain a general representation of the current shock ε_t as a linear combination of the present and past returns, with decaying weights θ^j. The decaying weighting pattern implies that the most remote returns have a smaller impact on current shocks compared with the most recent ones. The condition to be met to ensure that weights are decaying is $|\theta| < 1$. This is the invertibility condition for an MA(1) process.

In a similar way, we can show that the MA(q) process is invertible if all the roots of the polynomial $(1 + \theta_1 z + \theta_2 z^2 + \cdots + \theta_q z^q) = 0$ lie outside the unit circle.

2.3 Autoregressive moving-average (ARMA) processes

2.3.1 ARMA(1,1)

Most likely, neither an AR process nor an MA process is adequate to capture in full all the properties of a financial time series. Therefore, we now introduce a generalized model that combines the two types of models, the ARMA. We start by introducing the ARMA(1,1) model, where we have one lag for the AR part and one lag for the MA part. This model takes the following form:

$$r_t = \delta + \phi r_{t-1} + \theta \varepsilon_{t-1} + \varepsilon_t \tag{2.9}$$

where ε_t is a Gaussian white-noise process with a mean of 0 and a variance of σ^2. The process borrows properties from both AR and MA processes, with $|\phi| < 1$ being the stationarity condition and $|\theta| < 1$ being the invertibility condition. Using the lag polynomial $\theta(L) = (1 + \theta_1 L + \theta_2 L^2 + \cdots + \theta_q L^q)$ for the MA part and $\phi(L) = (1 - \phi_1 L - \phi_2 L^2 - \cdots - \phi_p L^p)$ for the AR part, we can rewrite (2.9) as

$$\phi(L) r_t = \theta(L) \varepsilon_t$$

The expected value of an ARMA(1,1) process is equal to a constant,

$$E(r_t) = E(\delta + \phi r_{t-1} + \theta \varepsilon_{t-1} + \varepsilon_t) \, E(r_t) = \frac{\delta}{1 - \phi}$$

and we can obtain the variance as follows:

$$\begin{aligned}
\text{var}(r_t) &= \text{var}(\delta + \phi r_{t-1} + \theta \varepsilon_{t-1} + \varepsilon_t) \\
\text{var}(r_t) &= \text{var}(\delta) + \phi^2 \text{var}(r_{t-1}) + \theta^2 \text{var}(\varepsilon_{t-1}) + \text{var}(\varepsilon_t) + 2\text{cov}(\delta, \phi r_{t-1}) \\
&\quad + 2\text{cov}(\delta, \theta \varepsilon_{t-1}) + 2\text{cov}(\delta, \varepsilon_t) + 2\text{cov}(\phi r_{t-1}, \theta \varepsilon_{t-1}) \\
&\quad + 2\text{cov}(\phi r_{t-1}, \varepsilon_t) + 2\text{cov}(\theta \varepsilon_{t-1}, \varepsilon_t) \\
\gamma_0 &= \phi^2 \gamma_0 + \theta^2 \sigma^2 + \sigma^2 + 2\phi\theta \, \text{cov}(r_{t-1}, \varepsilon_{t-1}) \\
\gamma_0 &= \phi^2 \gamma_0 + \theta^2 \sigma^2 + \sigma^2 + 2\phi\theta \, \text{cov}(\phi r_{t-2} + \theta \varepsilon_{t-2} + \varepsilon_{t-1}, \varepsilon_{t-1}) \\
\gamma_0 &= \phi^2 \gamma_0 + \theta^2 \sigma^2 + \sigma^2 + 2\phi\theta \, \text{var}(\varepsilon_{t-1}) \\
\gamma_0 &= \phi^2 \gamma_0 + \theta^2 \sigma^2 + \sigma^2 + 2\phi\theta \sigma^2 \\
\gamma_0 &= \phi^2 \gamma_0 + \sigma^2 (\theta^2 + 1 + 2\phi\theta) \\
\gamma_0 &= \frac{\sigma^2 (\theta^2 + 1 + 2\phi\theta)}{1 - \phi^2}
\end{aligned}$$

We skip the analytic derivation of the covariance of the ARMA(1,1) process and just report the theoretical AC:

$$\rho_0 = \frac{\gamma_0}{\gamma_0} = 1$$

$$\rho_1 = \frac{(\phi + \theta)(1 + \phi + \theta)}{1 - \phi^2}$$

$$\rho_k = \phi \rho_{k-1} \quad \text{for } k > 2$$

The behavior of the autocorrelation function depends on the sign of ϕ:

- If $\phi > 0$, then the ACF decreases exponentially, with the sign of ρ_1 equal to the sign of $(\phi + \theta)$.

- If $\phi < 0$, then the ACF decreases exponentially, changing signs, with the sign of ρ_1 equal to the sign of $(\phi + \theta)$.

To understand how an ARMA(1,1) process looks, we now simulate an ARMA(1,1) process with the code reported below:

```
clear all
set obs 1500

* fix the seed
set seed 1
generate time=_n
tsset time

* define parameters for ARMA(-0.8,0.3) simulation
scalar constant = 0.01
scalar AR = -0.8
scalar MA = 0.3
generate mynormal = rnormal()
generate y = .
replace y = mynormal in 1
replace y = constant + AR*L.y + MA*L.mynormal + mynormal if time>1
label variable y "ARMA(-0.8,0.3)"
```

We simulate two ARMA(1,1) processes with alternative values for the AR and the MA parameters. We report the simulated processes in figure 2.7.

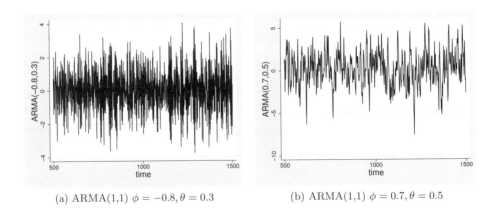

(a) ARMA(1,1) $\phi = -0.8, \theta = 0.3$ (b) ARMA(1,1) $\phi = 0.7, \theta = 0.5$

Figure 2.7. Simulations of ARMA(1,1) process

Similarly to what we did for both the AR and the MA processes, we attempt to identify a typical shape of the AC and the PAC for an ARMA(1,1) model by exploiting the theoretical AC just derived. In figure 2.8, we depict the AC and the PAC for an ARMA(1,1) with parameters $\phi = -0.8$ and $\theta = 0.3$.

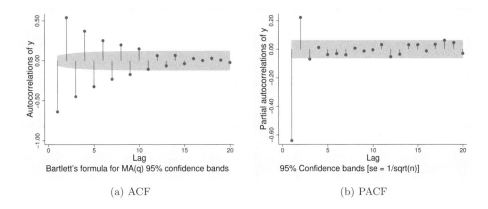

| (a) ACF | (b) PACF |

Figure 2.8. ACF and PACF for ARMA(1,1) process with parameters $(-0.8, 0.3)$

The AC decreases exponentially, changing sign from lag to lag, with the pattern being determined by the negative sign of the sum of the parameters $(\phi + \theta)$, which is equal to -0.5. The PAC decreases, changing sign and showing only two significant coefficients, with the first being negative.

Turning now to the ARMA(1,1) process with parameters $\phi = 0.7$ and $\theta = 0.5$, the AC and the PAC take the shapes reported in figure 2.9.

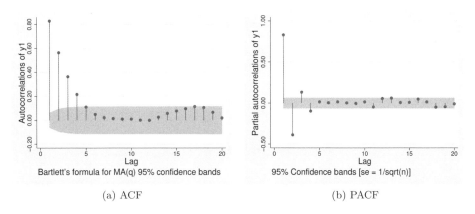

| (a) ACF | (b) PACF |

Figure 2.9. ACF and PACF for ARMA(1,1) process with parameters $(0.7, 0.5)$

The AC decreases exponentially, with positive sign, because the sign of the quantity $(\phi + \theta)$ suggests being equal to 1.2; the PAC decreases, changing sign, with only the first coefficients being statistically significant and the first taking a positive sign.

2.3.2   ARMA(p,q)

A generic ARMA(p, q) model is identified by p and q, the order of the AR and the MA components, respectively:

$$r_t = \delta + \sum_{i=1}^{p} \phi_i r_{t-i} + \sum_{j=1}^{q} \theta_j \varepsilon_{t-j} + \varepsilon_t$$

When working with financial data, a model with one lag for the AR part and one lag for the MA part will usually be sufficient to completely characterize the time series.

2.3.3   ARIMA

ARMA models describe stationary time series with some autocorrelation structure in the data. When the time series is not stationary, then we must move from an ARMA model to a more general ARIMA model representation, where "I" stands for the integrated order. An ARIMA model fits a unit-root nonstationary process.

From an empirical point of view, fitting an ARIMA model on the log-prices p_t is the same as fitting an ARMA model on first-differenced data (that is, on returns $p_t - p_{t-1}$). A time series y_t is said to follow an ARIMA$(p, 1, q)$ representation if the differenced series, $r_t = y_t - y_{t-1}$, follows a stationary and invertible ARMA(p, q) model.

We will revisit that point in section 2.4.

2.3.4   ARMAX

To improve the goodness of fit of the model, we may enrich our model by adding some extra variables that can help in describing the time pattern of the time series analyzed. This augmented ARMA is the so-called ARMAX model. The additional variables that we may use can be of any type. For instance, we could use macroeconomic variables, policy covariates, or variables capturing a peculiar phase of the time period analyzed, for instance, days with big drops or peaks or particular days of the week, such as Fridays. Again, we report some applications in section 2.4.

2.4   Application of ARMA models

We carry out our empirical analysis on the dataset introduced and discussed in chapter 1, that is, the daily returns of the S&P 500 for the years 1950–2013.

We stress that ARMA models are fit by maximum likelihood (ML). Assuming that returns r_t are conditionally normally distributed, then to fit an ARMA model, we have to maximize the following log-likelihood function:

$$\ln L\left(r_t; \vartheta\right) = -\frac{T}{2}\ln\left(2\pi\sigma^2\right) - \frac{1}{2}\sum_{t=2}^{T}\frac{\left\{r_t - \widehat{r}_t\left(\vartheta\right)\right\}^2}{\sigma^2}$$

where ϑ is the set of parameters defining the form of the ARMA process, $\widehat{r}_t(\vartheta)$ is the return estimate by the ARMA process, T is the length of the time series, and σ^2 is the variance.

When fitting a model under the assumption that returns are normally distributed, we should be aware about the properties of the estimates we are going to obtain according to the real distribution of the data on which we fit the model. An ML estimator is consistent if the assumptions about the true data-generating process imply that the expected value of the score equations is 0. Therefore, even when we are inappropriately assuming that returns follow a normal distribution provided we adopt robust standard errors, we can still obtain quasimaximum likelihood (QML) estimates that are consistent and asymptotically normal, especially when working with large samples. Robust standard errors are standard errors that are robust to some misspecification issues, for instance, when the data do not come from a simple random sample or the distribution of the data is not i.i.d.

In chapter 1, we showed that log-returns series are generally stationary, presenting some autocorrelation properties suggesting that an ARMA-type model could be successfully applied to our data. Therefore, we need to identify the ARMA model that best fits our data.

To this aim, we start by comparing the AC and the PAC of the empirical series with the theoretical ones for the alternative models considered in the previous sections. Remember that it is quite difficult to recognize the form of the process by just examining the form of the AC. Therefore, when fitting an ARMA model, it is meaningful to start with an overparameterized specification, for instance, including a certain number of lags for the AR part and only one for the MA part. We stop when all the coefficients are statistically significant and residuals follow a white-noise process, which implies that we have successfully modeled the historical time series.

Once we have fit a first model, we could see one of two cases:

1. Residuals do not show any long autocorrelation structure; therefore, we could try a simpler model, excluding some lags and starting from the coefficients that are not statistically significant. The purpose is to obtain a more parsimonious model.

2. The ACF of residuals shows some statistically significant autocorrelations; therefore, the model can be improved by adding some lags to the AR or to the MA part, possibly exploiting the indications coming from both the ACF and the PACF of residuals.

Recall that the information criteria we introduced in chapter 1 are overall good indicators that we can always use to choose between alternative models. We are going to use these widely in the following empirical sections.

We now start our example by recalling the ACF and the PACF for the S&P 500 by using the `ac` and the `pac` commands, as described before. We report the two charts in figure 2.10:

```
. use http://www.stata-press.com/data/feus/spdaily, clear
. tsset
. ac return, lags(40) name(AC)
. pac return, lags(40) name(PAC)
```

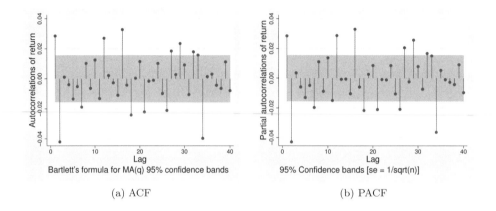

(a) ACF   (b) PACF

Figure 2.10. ACF and PACF for S&P daily returns

Unfortunately, neither the ACF nor the PACF give a clear idea about the proper number of terms characterizing the underlying model. However, we can identify a significant peak at the beginning of the sample together with some other significant peaks at lags 12 and 16. Therefore, a good starting model for this time series could be an ARMA(2,1).

With the purpose of illustrating the behavior of an extremely liquid time series, in figure 2.11 we report the ACF and the PACF for the daily euro–dollar exchange rate for the years 1999–2013.

```
. use http://www.stata-press.com/data/feus/eurusd
. generate time = _n
. tsset time
. generate delta = d.eurusd
. ac delta, name(AC)
. pac delta, name(PAC)
```

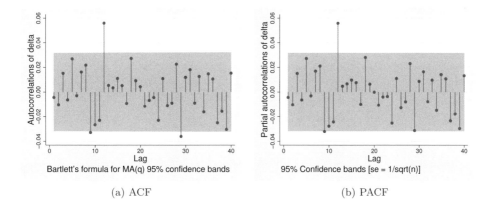

(a) ACF   (b) PACF

Figure 2.11. ACF and PACF for EUR–USD daily returns

Neither the ACF nor the PACF show any evidence of a time dependence structure in the data, with almost no peak being statistically significantly different from 0. This result is because of the high liquidity of the foreign market, making it extremely efficient.

2.4.1   Model estimation

We can use the `arima` command to fit ARMA models.

When checking the estimates, remember that Stata reports the intercept as the unconditional mean. For instance, given an ARMA(p, q) model,

$$r_t = \delta + \phi_1 r_{t-1} + \cdots + \phi_p r_{t-p} + \theta_1 \varepsilon_{t-1} + \cdots + \theta_q \varepsilon_{t-q} + \varepsilon_t$$

the intercept shown in the output is actually $\delta/1 - \phi_1 - \cdots - \phi_p$.

Before starting a full empirical implementation of an ARMA model, we briefly describe the estimation technique implemented in Stata. `arima` implements the conditional and the unconditional ML estimators. The conditional ML estimator drops the observations lost to lagged values of the dependent variable or lagged errors. The unconditional ML estimator uses the structure of the model to identify values to fill in for these missing values. The unconditional estimator can be more efficient and is frequently preferred. All the details can be found in [TS] **arima**.

As decided when we checked the ACF and the PACF, we now fit an ARMA(2,1) model on the S&P 500 daily returns:

```
. use http://www.stata-press.com/data/feus/spdaily, clear
. tsset newdate
        time variable:  newdate, 03jan1950 to 31dec2013
                delta:  1 day
```

```
. arima return, ar(1/2) ma(1)

(setting optimization to BHHH)
Iteration 0:    log likelihood =  51721.001
Iteration 1:    log likelihood =  51721.421
Iteration 2:    log likelihood =  51721.438
Iteration 3:    log likelihood =  51721.448
Iteration 4:    log likelihood =  51721.453
(switching optimization to BFGS)
Iteration 5:    log likelihood =  51721.457
Iteration 6:    log likelihood =  51721.465
Iteration 7:    log likelihood =  51721.467
Iteration 8:    log likelihood =  51721.469
Iteration 9:    log likelihood =  51721.469
Iteration 10:   log likelihood =  51721.469
Iteration 11:   log likelihood =  51721.469

ARIMA regression

Sample:  04jan1950 - 31dec2013            Number of obs     =      16102
                                          Wald chi2(3)      =     277.12
Log likelihood =  51721.47                Prob > chi2       =     0.0000
```

return	Coef.	OPG Std. Err.	z	P>|z|	[95% Conf. Interval]	
return						
_cons	.0002924	.0000799	3.66	0.000	.0001358	.0004491
ARMA						
ar						
L1.	-.068257	.0899525	-0.76	0.448	-.2445607	.1080466
L2.	-.0400998	.0041099	-9.76	0.000	-.0481552	-.0320445
ma						
L1.	.0981323	.0898873	1.09	0.275	-.0780435	.2743081
/sigma	.0097445	.0000152	640.54	0.000	.0097147	.0097743

```
Note: The test of the variance against zero is one sided, and the two-sided
      confidence interval is truncated at zero.
. estimates store ARMA21
```

We have fit a model with two lags for the AR part, `ar(2)`, and one lag for the MA part, `ma(1)`. Alternatively, we could have typed `arima returns, arima(2,0,1)`; here the first number indicates that we want to add two lags for the AR part, the second number indicates that we want to add the order of integration (here equal to 0), and the third number indicates that we want to add one lag to the MA part.

In the first part of the output, we find some information about the optimization procedure, with the iterations of the algorithm aimed at maximizing the log-likelihood function. The convergence is achieved in 11 steps, and it stops at the log-likelihood value of 51,721.47. We find this value just above the table. In addition, we are informed that the estimation sample consists of 16,102 observations and that the model is overall statistically significant, as suggested by the Wald test. The table provides parameters and standard errors, the t test for the statistical significance of parameters z and P>\|z\|, and the 95% confidence interval. `OPG Std. Err.` reminds us that Stata is using the outer product of the gradient to derive the estimates of the standard errors of the estimated coefficients. Finally, in the lower part of the table, we find the estimated standard deviation of the white-noise disturbance, `/sigma`, being equal to 0.0097, with a test for the null hypothesis that it is equal to 0.

Focusing now on the estimation results, only the second lag of the AR part is statistically significant, while both the AR(1) and the MA(1) do not contribute to explaining the time pattern of the process. Therefore, with the purpose of reducing the model and keeping just the significant variables, we remove the MA term.

```
. arima return, ar(1/2) nolog
ARIMA regression
Sample:  04jan1950 - 31dec2013                Number of obs   =       16102
                                              Wald chi2(2)    =      249.54
Log likelihood =  51721.35                    Prob > chi2     =      0.0000
```

		OPG				
return	Coef.	Std. Err.	z	P>|z|	[95% Conf.	Interval]
return						
_cons	.0002924	.0000794	3.68	0.000	.0001367	.0004481
ARMA						
ar						
L1.	.0296989	.0035393	8.39	0.000	.022762	.0366358
L2.	-.0429078	.0031311	-13.70	0.000	-.0490445	-.036771
/sigma	.0097446	.0000149	655.17	0.000	.0097154	.0097737

```
Note: The test of the variance against zero is one sided, and the two-sided
      confidence interval is truncated at zero.
. estimates store ARMA20
```

We specified the `ar(1/2)` option to indicate that we want only two lags for the AR part. Alternatively, we could have typed `arima(2,0,0)`. To hide the output of the iterative algorithm, we add the `nolog` option. The output above reports the estimates for the ARMA(2,0), where we can see that both AR coefficients turn out to be statistically significant, suggesting that this model specification should be preferred over the ARMA(2,1).

In addition, the log likelihood of the ARMA(2,1) is slightly larger than that of the ARMA(2,0), being equal to 51,721.47 and 51,721.35, respectively. This modest improvement does not justify the extra MA parameter in the ARMA(2,1) specification; therefore, we should prefer the ARMA(2,0) model. To formally provide evidence on that point, we compute information criteria for these two models:

```
. estimates stats ARMA21 ARMA20
```
Akaike´s information criterion and Bayesian information criterion

Model	Obs	ll(null)	ll(model)	df	AIC	BIC
ARMA21	16,102	.	51721.47	5	-103432.9	-103394.5
ARMA20	16,102	.	51721.35	4	-103434.7	-103403.9

Note: N=Obs used in calculating BIC; see [R] BIC note.

Both information criteria are lower for the ARMA(2,0) model, suggesting that this model should be preferred.

Having identified a possible model specification for our data, we now must check whether the model is correctly specified. The first step is to evaluate whether the residuals still present some autocorrelation patterns. To do this, we first have to store the residuals in a new variable by using the `predict` command. The purpose of `predict` is to obtain forecasts on the basis of the model just fit or to get residuals.

```
. predict residuals, residuals
(1 missing value generated)
```

We have stored the residuals from the fit model in a new variable, `residuals`. The name just after `predict` indicates the name of the new variable we are going to create, while the option after the comma indicates the type of prediction we want to obtain, in our case, `residuals`. For instance, if we want to obtain a linear prediction, we should specify `xb`.

We now draw the ACF of residuals in figure 2.12:

```
. ac residuals
```

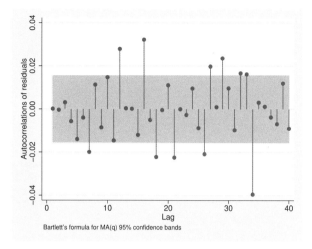

Figure 2.12. ACF for residuals of ARMA(2,0) model

Comparing the ACF reported in figure 2.10 with the one obtained on residuals in figure 2.12, we note that the evidence of two statistically significant peaks in the original time series disappears when considering residuals. This suggests that the ARMA(2,0) model has successfully modeled this feature of the data.

To choose among alternative models, we can use the information criteria with the likelihood-ratio test. For instance, although both coefficients are statistically significant in the ARMA(2,0) model, we may want to assess whether a more parsimonious model, say, an ARMA(1,0) model, can properly fit the data. We report below the estimates for the ARMA(1,0) model:

```
. arima return, ar(1)

  (output omitted)

ARIMA regression

Sample:  04jan1950 - 31dec2013              Number of obs   =        16102
                                            Wald chi2(1)    =        66.35
Log likelihood =  51706.51                  Prob > chi2     =       0.0000
```

return	Coef.	OPG Std. Err.	z	P>|z|	[95% Conf. Interval]
return					
_cons	.0002925	.0000817	3.58	0.000	.0001324 .0004526
ARMA					
ar					
L1.	.0284775	.003496	8.15	0.000	.0216255 .0353296
/sigma	.0097536	.0000149	656.71	0.000	.0097244 .0097827

```
Note: The test of the variance against zero is one sided, and the two-sided
      confidence interval is truncated at zero.
```

We now report in figure 2.13 the ACF of residuals from the ARMA(1,0) model:

```
. predict residualsAR1, residuals
. ac residualsAR1
```

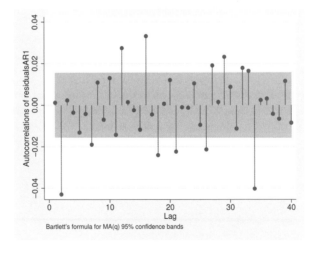

Figure 2.13. ACF for residuals of ARMA(1,0) model

Figure 2.13 shows that residuals of the ARMA(1,0) model are more strongly autocorrelated compared with residuals of the ARMA(2,0), as the ACF depicted in figure 2.12 suggests. This evidence would suggest that the ARMA(2,0) model outperforms the ARMA(1,0).

To carry out a proper comparison between the two models, we now adopt the information criteria as well as the likelihood-ratio test by using the `estat ic` command. We can obtain these indicators after each model, or, to obtain a nicer output, we can store the results with the `estimates store` command and then compare the two models by typing `estimates stats` followed by the name we used to store each model's results. We now illustrate the second approach:

```
. arima return, ar(1) nolog
  (output omitted )
. estimates store ARMA10
. estimates stats ARMA20 ARMA10
Akaike´s information criterion and Bayesian information criterion
```

Model	Obs	ll(null)	ll(model)	df	AIC	BIC
ARMA20	16,102	.	51721.35	4	-103434.7	-103403.9
ARMA10	16,102	.	51706.51	3	-103407	-103384

```
        Note: N=Obs used in calculating BIC; see [R] BIC note.
```

We have estimated an ARMA(2,0) model that we stored under the name **ARMA20** and an ARMA(1,0) that we stored under **ARMA10**.

The Akaike information criteria (AIC) takes a value equal to $-103,434.7$ for the ARMA(2,0) and to $-103,407$ for the ARMA(1,0), indicating that the ARMA(2,0) should be preferred. We get a similar indication from the Bayesian information criteria (BIC) because it is equal to $-103,403.9$ for the ARMA(2,0) and to $-103,384$ for the ARMA(1,0).

Instead of `estimates stats`, we can use the `estimates table` command followed by the names of the models we want to compare. We specify the `stats()` option, taking as an argument the names of the two information criteria, `aic` and `bic`.

```
. estimates table ARMA20 ARMA10, stats(aic bic)
```

Variable	ARMA20	ARMA10
return		
_cons	.00029241	.00029248
ARMA		
ar		
L1.	.02969892	.02847753
L2.	-.04290776	
sigma		
_cons	.00974457	.00975356
Statistics		
aic	-103434.69	-103407.02
bic	-103403.94	-103383.96

`estimates table` is useful because it allows us to quickly compare alternative models. The output reports the names of the models in columns and the parameters in rows. Thus, we can easily see that the AR(1) parameters change from 0.030 to 0.028 when moving from the ARMA(2,0) to the ARMA(1,0) specification.

Regarding the likelihood-ratio test, we need to compare the two models we have just stored as `ARMA20` and `ARMA10` by using the `lrtest` command.

```
. lrtest ARMA20 ARMA10
Likelihood-ratio test                             LR chi2(1)  =      29.67
(Assumption: ARMA10 nested in ARMA20)             Prob > chi2 =     0.0000
```

The ARMA(2,0) is the unrestricted model, and the ARMA(1,0) is the restricted model. The null hypothesis that we want to test by using the likelihood-ratio test is that the two models provide a similar fit, regardless of the number of parameters added into the model. The likelihood-ratio test strongly rejects the null hypothesis, suggesting that the ARMA(2,0) model provides a better fit to the data compared with the ARMA(1,0) model, because the extra parameter added in the ARMA(2,0) model is justified by the significant increase of the log likelihood. Note that we are informed that the assumption is that the `ARMA10` model is nested in the `ARMA20` model.

The indication from the likelihood-ratio test is consistent with the one from AIC and BIC: the best model is the ARMA(2,0).

We now show the effect of adding the `vce(robust)` option to the `arima` command, requiring standard errors to be computed using the robust version of the variance–covariance matrix. Fitting an ARMA model using robust standard errors allows us to move from the ML estimates to the QML estimates framework; that is, we can drop the normality assumption and obtain estimates that are still consistent and asymptotically normal.

```
. arima return, ar(1/2) vce(robust) nolog
ARIMA regression
Sample:  04jan1950 - 31dec2013              Number of obs   =       16102
                                            Wald chi2(2)    =        7.24
Log pseudolikelihood =  51721.35            Prob > chi2     =      0.0268
```

| | | Semirobust | | | | |
return	Coef.	Std. Err.	z	P>|z|	[95% Conf.	Interval]
return						
_cons	.0002924	.0000758	3.86	0.000	.0001439	.0004409
ARMA						
ar						
L1.	.0296989	.0184441	1.61	0.107	-.0064509	.0658488
L2.	-.0429078	.0200763	-2.14	0.033	-.0822566	-.0035589
/sigma	.0097446	.0002081	46.82	0.000	.0093366	.0101525

Note: The test of the variance against zero is one sided, and the two-sided
      confidence interval is truncated at zero.

```
. estimates store ARMA20robust
```

We can compare results with the case when standard errors are computed using the classical approach.

```
. estimates table ARMA20 ARMA20robust, se p
```

Variable	ARMA20	ARMA20ro~t
return		
_cons	.00029241	.00029241
	.00007942	.00007579
	0.0002	0.0001
ARMA		
ar		
L1.	.02969892	.02969892
	.00353931	.01844414
	0.0000	0.1074
L2.	-.04290776	-.04290776
	.00313107	.0200763
	0.0000	0.0326
sigma		
_cons	.00974457	.00974457
	.00001487	.00020815
	0.0000	0.0000

```
legend: b/se/p
```

The results reported above show that, when using the vce(robust) option, the estimated coefficients do not change. Meanwhile, standard errors get much higher, with the AR(1) moving from a level of statistical significance < 0.0001 to 0.11 and the AR(2) moving from < 0.0001 to 0.03. Robust standard errors are usually bigger than classical standard errors because of heteroskedasticity of the time series.

Just as we computed returns as simple first differences of log-prices, we now show that fitting an ARMA model on returns should provide the same results as fitting an ARIMA model on log-prices. Thus, we fit an ARIMA(2,1,0) model on log-prices.

```
. arima logprice, arima(2,1,0) nolog
ARIMA regression
Sample:  04jan1950 - 31dec2013                    Number of obs   =      16102
                                                  Wald chi2(2)    =     249.54
Log likelihood =  51721.35                        Prob > chi2     =     0.0000
```

| | | OPG | | | | |
D.logprice	Coef.	Std. Err.	z	P>|z|	[95% Conf.	Interval]
logprice						
_cons	.0002924	.0000794	3.68	0.000	.0001367	.0004481
ARMA						
ar						
L1.	.0296989	.0035393	8.39	0.000	.022762	.0366358
L2.	-.0429078	.0031311	-13.70	0.000	-.0490445	-.036771
/sigma	.0097446	.0000149	655.17	0.000	.0097154	.0097737

```
Note: The test of the variance against zero is one sided, and the two-sided
      confidence interval is truncated at zero.
```

As expected, the estimated coefficients are exactly the same as those obtained by fitting an ARMA(2,0) model on returns.

2.4.2 Postestimation

As already discussed in the previous section, a first step to assess whether the model is correctly specified is to analyze the autocorrelation structure of the residuals. If the model is correctly specified, then residuals should follow a white-noise process; if residuals show some time pattern, then the model is underparameterized.

Several postestimation commands can help us evaluate whether the model is correctly specified.

Starting from the analysis of residuals, we can use the wntestq command, which performs the Portmanteau (or Q) test for white noise. The null hypothesis is that the residuals are distributed like a white-noise process. To show how this command works, let us consider our ARMA(2,0) model and perform the test on residuals for this model:

```
. arima return, ar(1/2) nolog
  (output omitted )
. predict residualsAR2, residual
(1 missing value generated)
. wntestq residualsAR2
Portmanteau test for white noise

 Portmanteau (Q) statistic =    137.6852
 Prob > chi2(40)           =      0.0000
```

The results reported above show that the null hypothesis that residuals follow a white-noise process is strongly rejected; therefore, the ARMA(2,0) model cannot completely describe the structure of the data. This indication is coherent with figure 2.10, where we noticed that some lags were statistically significant.

In addition to the Portmanteau test, we can draw the Bartlett's periodogram evaluating the null hypothesis that the data come from a white-noise process of uncorrelated random variables, with a constant mean and a constant variance. We can use the `wntestb` command followed by the name of the variable we want to evaluate, in our case, `residualsAR2`:

```
. wntestb residualsAR2
```

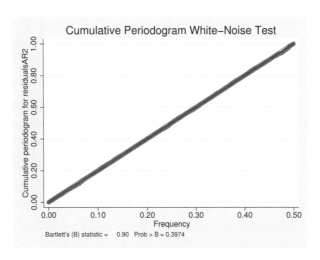

Figure 2.14. Bartlett's periodogram for residuals of ARMA(2,0) model

The values lie inside the confidence intervals, suggesting that residuals follow a white-noise process. In addition to the graphical representation, in the lower part of the figure we find the test statistics. The `wntestb` command even allows us to obtain a formal test with specification of the `table` option:

```
. wntestb residualsAR2, table
Cumulative periodogram white-noise test
```

Bartlett´s (B) statistic	=	0.8966
Prob > B	=	0.3974

The *p*-value is equal to 0.40, indicating that the process characterizing residuals is not different from a white-noise process, leading therefore to a different conclusion than the `wntestq` command.

In addition to these commands aimed at evaluating the distribution of residuals, we can explore other useful tools after fitting an ARMA model. For instance, the `estat acplot` command draws the estimated autocorrelation and autocovariance functions for a stationary process by using the parameters of a fitted model. For our ARMA(2,0) model, we obtain figure 2.15:

```
. estat acplot
```

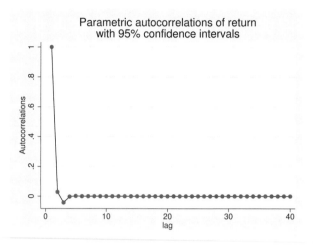

Figure 2.15. ACF for ARMA(2,0) model

From figure 2.15, we cannot infer about the form of the model. In contrast, considering a simulated AR(1) model, we get a clear indication about the model specification, as depicted in figure 2.16:

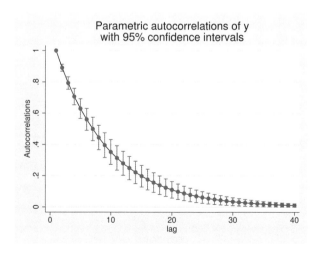

Figure 2.16. ACF for simulated data, AR(1)

We can carry out additional postestimation evaluations aimed at assessing the stability of solutions of the companion matrices for the AR and MA polynomials that corresponds to the stability of eigenvalues. We can use the `estat aroots` command to generate a graph of the eigenvalues.

```
. use http://www.stata-press.com/data/feus/spdaily, clear
. tsset newdate
        time variable:  newdate, 03jan1950 to 31dec2013
                delta:  1 day
. arima return, ar(1/2) nolog
  (output omitted )
. estimates store ARMA20
. estat aroots
```

Eigenvalue stability condition

Eigenvalue		Modulus
.01484946 +	.2066089*i*	.207142
.01484946 −	.2066089*i*	.207142

All the eigenvalues lie inside the unit circle.
AR parameters satisfy stability condition.

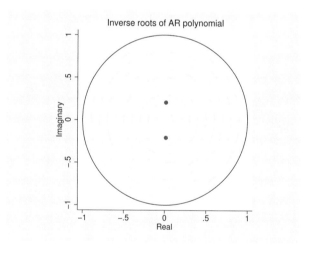

Figure 2.17. Eigenvalues of companion matrix of AR polynomial

As described in section 2.1, an ARMA model is stationary if the modulus of each eigenvalue of the companion matrix is strictly less than 1. The graph reported in figure 2.17 shows the eigenvalues of the companion matrices for the AR and MA polynomials, which are the inverse roots of the AR polynomial in (2.5). From a graphical point of view, the stability condition is met if the eigenvalues lie inside the unit circle. If this condition holds true, the process is stationary, invertible, and has an infinite-order

MA representation. Figure 2.17 displays the eigenvalues with the real components on the x axis and the imaginary components on the y axis. For the case of our ARMA(2,0) model, the eigenvalues lie inside the unit circle; therefore, we can conclude that our model satisfies the stability condition.

The last command we would like to illustrate is the impulse–response function, `irf`. We can estimate and plot (see figure 2.18) the impulse–response function on the ARMA(2,0) model as follows:

```
. irf create AR2, set(myirf1, replace)
(file myirf1.irf created)
(file myirf1.irf now active)
(file myirf1.irf updated)

. irf graph irf
```

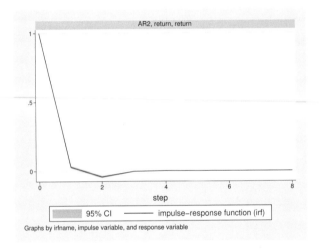

Figure 2.18. Impulse–response function for ARMA(2,0) model

The impulse–response function shows that a positive shock initially causes an increase in returns, but the persistence of this shock is limited because its impact almost immediately dies out, with the response function approaching 0 starting at the first step.

We can obtain a table with the result of the impulse–response function by using the `irf table` command:

```
. irf table irf
             Results from AR2
```

	(1)	(1)	(1)
step	irf	Lower	Upper
0	1	1	1
1	.029699	.022762	.036636
2	-.042026	-.04816	-.035891
3	-.002522	-.003217	-.001828
4	.001728	.001217	.002239
5	.00016	.0001	.000219
6	-.000069	-.000101	-.000038
7	-8.9e-06	-.000013	-4.5e-06
8	2.7e-06	1.1e-06	4.4e-06

```
95% lower and upper bounds reported
(1) irfname = AR2, impulse = return, and response = return
```

The rows of the tables indicate the time since impulse, while each column reports the value of the impulse–response function, `irf`, with a confidence level, `Lower` and `Upper`.

The `irf graph` and `irf table` commands allow us to obtain more sophisticated response functions, such as the orthogonalized impulse–response function, the dynamic-multiplier function, and the cumulative impulse–response function. See [TS] **irf** for further information.

2.4.3 Adding a dummy variable

From a simple descriptive analysis of the S&P returns, there is evidence of the presence of a big outlier corresponding to 19 October 1987, the Black Monday. We now investigate the impact that a single observation can have on the entire model estimation. We proceed by adding a dummy variable taking a value of 1 for Black Monday and a value of 0 otherwise. Formally, when augmenting the model by an extra variable, we are fitting an ARMAX model. We use the code reported below:

```
. generate dummy = 0
. replace dummy = 1 if date==td(19Oct1987)
(1 real change made)
```

Our first step is to create a variable with name `dummy` by using the `generate` command and then initialize it to 0. The second step is to set the `dummy` variable to 1 for Black Monday by using the `replace` command. More synthetically, we could use the following command:

```
. generate dummybis = (date==td(19Oct1987))
```

We now refit the ARMA(2,0) specification by adding the dummy variable we just generated. The results are reported below:

```
. arima return dummy, ar(1/2) nolog

ARIMA regression

Sample:  04jan1950 - 31dec2013                Number of obs    =        16102
                                              Wald chi2(3)     =       202.74
Log likelihood =  51997.56                    Prob > chi2      =       0.0000
```

	Coef.	OPG Std. Err.	z	P>|z|	[95% Conf.	Interval]
return						
dummy	-.2270724	.0260273	-8.72	0.000	-.278085	-.1760599
_cons	.0003065	.0000773	3.96	0.000	.000155	.0004581
ARMA						
ar						
L1.	.0302587	.0042858	7.06	0.000	.0218588	.0386587
L2.	-.0348128	.0037935	-9.18	0.000	-.042248	-.0273776
/sigma	.0095788	.0000231	413.82	0.000	.0095335	.0096242

```
Note: The test of the variance against zero is one sided, and the two-sided
      confidence interval is truncated at zero.

. estimates store ARMA20Dummy
```

The dummy variable is statistically significant, and it has a coefficient quite close to the return observed on Black Monday, minus the constant. The two AR parameters are quite different from those we obtained when fitting the ARMA(2,0) model without adding the dummy variable. This finding indicates how big the impact of a single outlier can be.

From a mathematical point of view, fitting the model by adding a dummy variable corresponding to Black Monday provides the same results as if we had fit the model without that observation:

```
. arima return if date!=td(19Oct1987), ar(1/2) nolog
Number of gaps in sample:  1
(note: filtering over missing observations)
ARIMA regression
Sample:  04jan1950 - 31dec2013, but with a gap   Number of obs   =      16101
                                                 Wald chi2(2)    =     136.20
Log likelihood =  51993.83                       Prob > chi2     =     0.0000
```

		OPG				
return	Coef.	Std. Err.	z	P>|z|	[95% Conf. Interval]	
return						
_cons	.0003065	.000077	3.98	0.000	.0001555	.0004575
ARMA						
ar						
L1.	.0302563	.0042093	7.19	0.000	.0220061	.0385064
L2.	-.0348079	.0037838	-9.20	0.000	-.042224	-.0273919
/sigma	.0095791	.0000228	419.24	0.000	.0095343	.0096239

Note: The test of the variance against zero is one sided, and the two-sided
      confidence interval is truncated at zero.

In this case, we exclude the return observed on Black Monday by requiring that the date be different from 19 October 1987, `if date!=td(19Oct1987)`. In fact, Stata tells us that the sample has a gap: `Sample: 04jan1950 - 31dec2013, but with a gap`.

We conclude the analysis of the impact of Black Monday on the overall sample by comparing the two models, ARMA(2,0) and ARMA(2,0) plus the dummy variable, via the information criteria and the likelihood-ratio test.

```
. estimates table ARMA20 ARMA20Dummy, stats(aic bic)
```

Variable	ARMA20	ARMA20Du~y
return		
dummy		-.22707245
_cons	.00029241	.00030653
ARMA		
ar		
L1.	.02969892	.03025874
L2.	-.04290776	-.03481277
sigma		
_cons	.00974457	.00957882
Statistics		
aic	-103434.69	-103985.12
bic	-103403.94	-103946.68

The information criteria show that the ARMA(2,0) augmented by the dummy variable should be preferred because both AIC and BIC take a smaller value.

```
. lrtest ARMA20 ARMA20Dummy

Likelihood-ratio test                                    LR chi2(1)  =     552.43
(Assumption: ARMA20 nested in ARMA20Dummy)               Prob > chi2 =     0.0000
```

The likelihood-ratio test confirms that adding the Black Monday dummy variable to the ARMA(2,0) model helps in improving the model goodness of fit.

2.4.4 Forecasting

Sometimes, time-series analysis aims at exploiting current information to draw some inference about future values, which means making forecasts. The general setup is that we are at time t, called the forecast origin, and we want to obtain a prevision at time $t + k$, that is, r_{t+k}, where $k \geq 1$ is the forecast horizon.

Given an ARMA(1,1) process, we have

$$r_t = \delta + \phi r_{t-1} + \theta \varepsilon_{t-1} + \varepsilon_t$$

Given the information set I_t, at the forecast horizon $k = 1$, we have

$$E\left(r_{t+1}|I_t\right) = E\left(\delta + \phi r_t + \theta \varepsilon_t + \varepsilon_{t+1}|I_t\right)$$
$$E\left(r_{t+1}|I_t\right) = \delta + \phi r_t + \theta \varepsilon_t$$

and at horizon $k = 2$, we have

$$E\left(r_{t+2}|I_t\right) = E\left(\delta + \phi r_{t+1} + \theta \varepsilon_{t+1} + \varepsilon_{t+2}|I_t\right)$$
$$E\left(r_{t+2}|I_t\right) = \delta + \phi E\left(r_{t+1}|I_t\right)$$
$$E\left(r_{t+2}|I_t\right) = \delta + \phi\left(\delta + \phi r_t + \theta \varepsilon_t\right)$$
$$E\left(r_{t+2}|I_t\right) = \delta\left(1 + \phi\right) + \phi^2 r_t + \phi \theta \varepsilon_t$$

Generalizing, at time horizon k, the forecast is

$$E\left(r_{t+k}|I_t\right) = \delta\left(1 + \phi\right) + \phi^k r_t + \phi^{k-1} \theta \varepsilon_t$$

Under the stationarity condition $|\phi| < 1$ and for $k \to \infty$, the forecast tends to the unconditional mean of the series $\delta/1 - \phi$. We can therefore conclude that forecasts based on an ARMA(1,1) process are valid only for the short horizon. This result is coherent with the efficient market hypothesis.

The basic command to obtain forecasts is `predict`, which we have already used for storing residuals.

To obtain one-step forecasts for the S&P 500 based on the ARMA(2,0) model, after fitting the model, we use the `predict forecasts, xb` command. We graph the prediction in figure 2.19 represented by `tsline return forecasts`:

```
arima return, ar(1/2) nolog
predict forecasts, xb
tsset date
tsline return forecasts
```

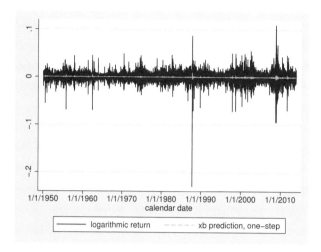

Figure 2.19. S&P daily in-sample forecasts based on ARMA(2,0)

From figure 2.19, we can see that forecasts based on the ARMA(2,0) model are quite poor, because they are roughly constant through time. This result is in line with the efficient market hypothesis and with the assumption of prices to follow a martingale process, both formulated at the beginning of chapter 1.

3 Modeling volatilities, ARCH models, and GARCH models

3.1 Introduction

Volatility measures the variation of a financial time series over time. Volatility can be estimated in several ways, with the simplest being the historical volatility defined as the standard deviation of past realizations of the financial time series.

Empirical evidence shows that volatility today depends on volatility yesterday; that is, days with high volatility are followed by days with high volatility and days with low volatility tend to be followed by days with low volatility. Therefore, the ARMA models we discussed in chapter 2 will constitute the baseline for the volatility models we introduce in this chapter.

From an economic point of view, we can explain volatility clustering by three pieces of empirical evidence:

1. Pieces of information arrive together, and some of them have a greater impact than others, for example, macroeconomic data.

2. Because of differences in trading times, negotiation can be seen as a continuous process. This is why there can be a collection of information coming from other markets right when a market opens, for instance, information on the European economy becomes available to the U.S. market at its opening.

3. There are psychological effects due to the arrival of important information, shaping a sentiment of uncertainty about the economic climate. In a nervous environment, the impact of news is amplified.

We can evaluate the presence of volatility clustering with the half-life, a measure of how fast the forecast of the conditional volatility h_t converges to the unconditional volatility σ^2, where the definition of conditional and unconditional volatility depends on whether volatility is computed conditioning on past information. From a mathematical point of view, the half-life is equal to the time necessary to halve the distance between the conditional variance h_t and the unconditional variance σ^2:

$$h_{t+k|t} - \sigma^2 = \frac{1}{2}\left|h_{t+1|t} - \sigma^2\right|$$

To provide a visual inspection of heteroskedasticity, we report the plot of squared returns for the daily S&P 500 returns time series (see figure 3.1):

```
use http://www.stata-press.com/data/feus/spdaily
generate return2 = return^2
tsset date
tsline return2 if date != td(19Oct1987)
```

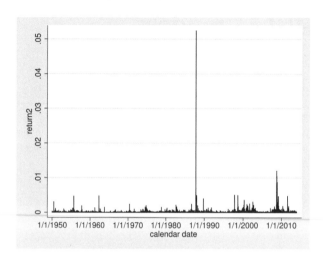

Figure 3.1. Squared daily S&P 500 returns

The figure shows periods of low volatility followed by periods of higher volatility, synonymous of a clear autoregressive nature of the volatility of asset returns. The huge spike corresponds to Black Monday.

Useful references for the material in this chapter are Bauwens, Hafter, and Laurent (2011, 2012), Engle and Granger (2003), Bollerslev (2010), Teräsvirta (2009), and Zivot (2009).

3.2   ARCH models

Autoregressive conditional heteroskedasticity (ARCH) models were introduced by Engle (1982) to model volatility. This class of models is motivated by the strong empirical evidence that returns are heteroskedastic.

In chapter 2, we introduced ARMA models, aimed at describing the behavior of the level of asset returns. We showed that a reasonable model we can assume for returns is the AR(1):

$$r_t = \delta + \phi r_{t-1} + \varepsilon_t$$

where ε_t is the innovation process, following a white noise with a mean of 0 and a constant variance of σ^2. Innovations, such as macroeconomic news releases, can be interpreted as news. In figure 3.1, we show that the hypothesis that the variance σ^2 is constant does not hold true because of heteroskedasticity.

Let us introduce the ARCH(1) model. According to Engle (1982), news is supposed to follow the product process,

$$\varepsilon_t = \eta_t \sqrt{h_t} \tag{3.1}$$

where $\eta_t | I_{t-1}$ is an i.i.d. process following a standard normal distribution, while I_{t-1} is the information set available up to time $t-1$. The notation $|$ indicates that we are going to consider the distribution of η_t conditioned on the information set I_{t-1}.

In addition, we have

$$h_t = \omega + \alpha \varepsilon_{t-1}^2$$
$$\varepsilon_t | I_{t-1} \approx N(0, h_t)$$
$$E(\eta_t, \varepsilon_t) = 0$$

According to the ARCH model, innovations ε_t are no longer following a distribution with constant variance σ^2, but now their conditional distribution $\varepsilon_t | I_{t-1}$ has a time-dependent variance h_t, being modeled as $h_t = \omega + \alpha \varepsilon_{t-1}^2$, where ω is a constant. The heteroskedasticity feature of the returns data is captured by letting the conditional variance at time t, h_t, be dependent on innovations at time $t-1$, ε_{t-1}.

We can compute the expected value of the process, assuming that $\varepsilon_t | I_{t-1} \simeq N(0, h_t)$, as follows:

$$E\left(\varepsilon_t\right) = E\left(\eta_t \sqrt{h_t}\right)$$
$$E\left(\varepsilon_t\right) = E\left(\eta_t \sqrt{\omega + \alpha \varepsilon_{t-1}^2}\right)$$
$$E\left(\varepsilon_t\right) = E\left(\eta_t\right) E\left(\sqrt{\omega + \alpha \varepsilon_{t-1}^2}\right)$$
$$E\left(\varepsilon_t\right) = 0$$

We can obtain the unconditional or long-run variance as

$$\sigma^2 = \text{Var}(\varepsilon_t) = E\left\{\varepsilon_t - E\left(\varepsilon_t\right)\right\}^2$$
$$\sigma^2 = E\left(\varepsilon_t\right)^2$$
$$\sigma^2 = E\left\{\eta_t^2\left(\omega + \alpha \varepsilon_{t-1}^2\right)\right\}$$
$$\sigma^2 = E\left(\eta^2 \omega\right) + E\left(\eta_t^2 \alpha \varepsilon_{t-1}^2\right)$$
$$\sigma^2 = \omega + \alpha E\left(\varepsilon_{t-1}^2\right)$$
$$\sigma^2 = \omega + \alpha \sigma^2$$

allowing us to rewrite the unconditional variance as

$$\sigma^2 = \omega + \alpha \sigma^2 = \frac{\omega}{1 - \alpha} \tag{3.2}$$

which is well defined if $\omega > 0$ and $0 \leq \alpha < 1$.

Going forward, we can show that the ARCH model effectively considers innovations to be independently distributed (remember that under the normality assumption, absence of correlation implies independence):

$$
\begin{aligned}
\text{Cov}(\varepsilon_t, \varepsilon_{t-1}) &= E\left(\eta_t \sqrt{h_t} \eta_{t-1} \sqrt{h_{t-1}}\right) \\
\text{Cov}(\varepsilon_t, \varepsilon_{t-1}) &= E\left(\eta_t \sqrt{\omega + \alpha \varepsilon_{t-1}^2} \, \eta_{t-1} \sqrt{\omega + \alpha \varepsilon_{t-2}^2}\right) \\
\text{Cov}(\varepsilon_t, \varepsilon_{t-1}) &= E\left(\eta_t\right) E\left(\eta_{t-1}\right) E\left(\sqrt{\omega + \alpha \varepsilon_{t-1}^2} \sqrt{\omega + \alpha \varepsilon_{t-2}^2}\right) \\
\text{Cov}(\varepsilon_t, \varepsilon_{t-1}) &= 0
\end{aligned}
$$

Finally, we can explicitly obtain the expression of the conditional variance at time t as follows:

$$
\begin{aligned}
h_t &= \text{Var}\left(\varepsilon_t | I_{t-1}\right) \\
h_t &= E\left[\left\{\eta_t^2 \left(\omega + \alpha \varepsilon_{t-1}^2\right)\right\} | I_{t-1}\right] \\
h_t &= E\left(\eta_t^2 \omega | I_{t-1}\right) + E\left(\eta_t^2 \alpha \varepsilon_{t-1}^2 | I_{t-1}\right) \\
h_t &= \omega + \alpha \varepsilon_{t-1}^2
\end{aligned}
$$

According to ARCH models, the conditional variance h_t is mean-reverting toward the long-run unconditional variance σ^2. We therefore now show that the expected value of the conditional variance is equal to the unconditional variance:

$$
\begin{aligned}
E\left(h_t\right) &= E\left\{\text{Var}\left(\varepsilon_t | I_{t-1}\right)\right\} \\
E\left(h_t\right) &= E\left(\varepsilon_t^2\right) \\
E\left(h_t\right) &= \sigma^2
\end{aligned}
$$

We can appreciate the mean-reverting property of the conditional volatility by substituting for the constant in (3.2) the unconditional variance σ^2 in the conditional variance equation:

$$
\begin{aligned}
h_t &= \omega + \alpha \varepsilon_{t-1}^2 \\
h_t &= \sigma^2 \left(1 - \alpha\right) + \alpha \varepsilon_{t-1}^2 \\
h_t &= \sigma^2 + \alpha \left(\varepsilon_{t-1}^2 - \sigma^2\right)
\end{aligned}
\tag{3.3}
$$

From (3.3), it is clear that we can express the conditional volatility h_t as the sum of the unconditional variance σ^2 and of the difference between the most recent squared innovation ε_{t-1}^2 and its expected value σ^2 (this quantity represents the short-term deviation from the long-run variance due to the surprise effect brought to the market at time t).

We can fit ARCH models by using the **arch** command. The **arch** command allows many options; some of them referring to the kind of the model we are working with being model-specific, for instance, just for ARCH or GARCH models, and some of them being more general, for instance, affecting the way standard errors are computed or the way the maximization procedure is carried out. We will start from describing this second set of options, because they are required for understanding how to estimate and evaluate the goodness of fit of a simple ARCH model, to move afterward to describe all the available ARCH models implemented in Stata.

3.2.1 General options

The general options we now describe for the `arch` command are the following:

- `noconstant` suppresses the constant term.

- `arch(`*numlist*`)` specifies the terms ARCH should include. *numlist* may be specified in several ways; for instance, for testing for lags 1 to n, we may type (1 2 3 ... n), (1/n), (1 n), (1 2 to n), or (1 2:n).

- `distribution(dist` $\left[\,\#\,\right]$`)` specifies which distribution to use among the normal (`gaussian` or `normal`), the Student's t (`t`), or the generalized error distribution (`ged`). The default is the normal distribution.

ARCH

We start by fitting the simplest model, an ARCH model with just one lag and under the assumption of innovations following a normal distribution. We have to specify that we want to add just one lag for the ARCH part, `arch(1)`.

```
. tsset newdate
        time variable:  newdate, 03jan1950 to 31dec2013
                delta:  1 day
. arch return, arch(1) nolog

ARCH family regression
Sample: 04jan1950 - 31dec2013                    Number of obs   =       16,102
Distribution: Gaussian                           Wald chi2(.)    =            .
Log likelihood =   52723.7                       Prob > chi2     =            .
```

	Coef.	OPG Std. Err.	z	P>|z|	[95% Conf. Interval]	
return						
_cons	.0004271	.0000603	7.09	0.000	.000309	.0005453
ARCH						
arch						
L1.	.3165625	.0066924	47.30	0.000	.3034457	.3296793
_cons	.0000654	4.35e-07	150.39	0.000	.0000645	.0000662

```
. estimates store ARCH
```

The upper part of the output provides us with some information concerning the estimation. We see that the sample size is composed of 16,102 observations; the first one was dropped to properly initialize the model. The number of observations dropped is the same as the number of p lags of the ARCH model. The output also reminds us that we are working under the assumption that returns are normally distributed. Just above the table containing the estimates, we find the log likelihood that takes a value equal to 52,723.7, which is the value where the iteration procedure stopped. Finally, we are provided with a test about the statistical significance of the model for the mean equation, which is the Wald χ^2 test and its corresponding p-value. In our case, we are just adding the constant term in the mean equation, making the test of the specification of the mean equation meaningless, which is why the values of both the Wald test and the p-value are missing.

We can see that the table reporting the estimates is divided into two parts, with the upper part showing the estimates for the mean equation, which in this case consists of just the constant, while the lower part is about the volatility equation, with just one lag for the ARCH part. The estimates for both the mean and the variance equations are statistically different from 0.

Distribution

Choosing the proper distribution is fundamental to get more efficient point estimates and to better accommodate the extreme errors that financial markets produce.

We have just fit our ARCH(1) model under the assumption that returns are normally distributed. However, when we checked the Gaussianity assumption for the S&P 500 returns in chapter 1, we showed that they are far from being normally distributed.

In this respect, Weiss (1986) and Bollerslev and Wooldridge (1992) show that even when normality is inappropriately assumed, maximizing the Gaussian log-likelihood results in QML estimation that is consistent and asymptotically normally distributed, although inefficient, provided the conditional mean and variance functions of the GARCH model are correctly specified. The implication for our empirical application is that even when errors are not normally distributed, the QML estimates (which we can obtain by specifying `arch` with normal errors and the `vce(robust)` option) are consistent and asymptotically normal, and that the variance–covariance of the estimator (VCE) is consistently estimated.

Engle and Gonzalez-Rivera (1991) and Bollerslev and Wooldridge (1992) focus on evaluating the accuracy of the QML estimation of GARCH(1,1) models, showing that if the distribution of standardized residuals is symmetric, then QML estimation is often close to ML estimation. However, if standardized residuals follow a skewed distribution, then the QML estimation can be quite different from the ML estimation. Finally, assuming the fourth-order moment exists, Bollerslev (1986) showed that the kurtosis implied by a GARCH(1,1) process with normal errors is greater than 3. This result implies that even when we are fitting a GARCH model under the Gaussianity assumption, we can still capture the presence of fat tails. However, to properly model fat tails and asymmetry, we have to fit a GARCH model under the assumption of nonnormal distribution.

Assuming that returns are conditionally normally distributed, the estimation of an ARCH model is carried out by maximizing the following log-likelihood function:

$$\ln L = -\frac{T}{2} \ln (2\pi h_t) - \frac{1}{2} \sum_{t=2}^{T} \frac{\varepsilon_t^2}{h_t} \tag{3.4}$$

where T is the length of the time series, h_t is the conditional time-varying variance, and ε_t is the innovation process.

To illustrate how the vce(robust) option works, we now fit an ARCH(1) model under the normality assumption using this option, and we compare results with the previously fit model without correction:

```
. arch return, arch(1) vce(robust) nolog

ARCH family regression

Sample: 04jan1950 - 31dec2013                    Number of obs   =     16,102
Distribution: Gaussian                           Wald chi2(.)    =          .
Log pseudolikelihood =    52723.7                Prob > chi2     =          .
```

	Coef.	Semirobust Std. Err.	z	P>|z|	[95% Conf. Interval]
return						
_cons	.0004271	.000083	5.14	0.000	.0002644	.0005898
ARCH						
arch						
L1.	.3165625	.0341326	9.27	0.000	.2496639	.3834612
_cons	.0000654	2.24e-06	29.23	0.000	.000061	.0000698

```
. estimates store ARCHRobust

. estimates table ARCH ARCHRobust, se
```

Variable	ARCH	ARCHRobust
return		
_cons	.00042713	.00042713
	.00006028	.00008302
ARCH		
arch		
L1.	.31656254	.31656254
	.00669236	.03413258
_cons	.00006537	.00006537
	4.347e-07	2.237e-06

legend: b/se

The point estimates of the two models are identical, but standard errors are quite different, with those obtained from the robust version being much larger with respect to the case where the correction is not used.

When fitting ARCH models, we are provided with two alternative distributions to the normal one, namely, the Student's t and the generalized error distribution (GED). When we are working with fat-tail returns, we can think about using one of these two distributions.

Under the assumption that the conditional distribution of ε_t follows a Student's t distribution with v degrees of freedom, the log-likelihood function for the ARCH model takes the following form:

$$\ln L = T \ln \left\{ \frac{\Gamma\left(\frac{v+1}{2}\right)}{\sqrt{\pi(v-2)}\Gamma\left(\frac{v}{2}\right)} \right\} - \frac{1}{2}\sum_{t=1}^{T}\ln h_t - \frac{v+1}{2}\sum_{t=1}^{T}\ln\left\{ 1 + \frac{\varepsilon_t^2}{h_t(v-2)} \right\} \quad (3.5)$$

where $\Gamma(\cdot)$ denotes the gamma function and v is the degrees of freedom satisfying the condition $v > 2$.

We now consider an ARCH(1) model under the assumption of a Student's t distribution. When we carried out the descriptive analysis of the S&P 500 returns in chapter 1, we found that the kurtosis was 30.69, suggesting that the normal distribution could not properly describe this financial time series. We could therefore reasonably expect that a fat-tail distribution might better fit our data. To fit an ARCH(1) model with a Student's t distribution, we use the arch command with the distribution(t) option:

```
. arch return, arch(1) distribution(t) nolog
ARCH family regression
Sample: 04jan1950 - 31dec2013                    Number of obs   =     16,102
Distribution: t                                  Wald chi2(.)    =          .
Log likelihood =  54105.25                       Prob > chi2     =          .
```

	Coef.	OPG Std. Err.	z	P>|z|	[95% Conf. Interval]	
return						
return						
_cons	.0005256	.0000579	9.08	0.000	.0004122	.000639
ARCH						
arch						
L1.	.3246211	.0218095	14.88	0.000	.2818754	.3673669
_cons	.0000689	2.28e-06	30.25	0.000	.0000644	.0000733
/lndfm2	.4900573	.072327	6.78	0.000	.3482989	.6318157
df	3.63241	.1180674			3.416656	3.881023

```
. estimates store ARCH1T
```

Just above the table in the output, we are reminded that we are now working under the assumption that returns follow the Student's t distribution. Comparing these results with those of the previous output about an ARCH model with Gaussian distribution, here we can find an extra parameter, df, indicating the number of degrees of freedom of a Student's t distribution and corresponding to the parameter v in (3.5). The degrees of freedom provide information about the degree of thickness of the tails of the distribution: the lower the degrees of freedom, the thicker the tails (and vice versa). The Student's *t* distribution approaches the normal one when the degrees of freedom gets larger than 30. Here the degrees of freedom is equal to 3.63, indicating that the error distribution is characterized by very fat tails.

We now compare the goodness of fit of the ARCH(1) model fit under the normal distribution and under the Student's *t* distribution.

```
. estimates stats ARCH ARCH1T
Akaike's information criterion and Bayesian information criterion
```

Model	Obs	ll(null)	ll(model)	df	AIC	BIC
ARCH	16,102	.	52723.7	3	-105441.4	-105418.3
ARCH1T	16,102	.	54105.25	4	-108202.5	-108171.7

Note: N=Obs used in calculating BIC; see [R] BIC note.

The AIC is equal to $-105{,}441.4$ for the ARCH model with normal innovations and equal to $-108{,}202.5$ for the ARCH model with Student's *t* innovations. The BIC is equal to $-105{,}418.3$ and $-108{,}171.7$, respectively. Therefore, both criteria clearly indicate that the Student's *t* distribution provides a better fit of the data compared with the Gaussian one. Note that here we cannot directly compare the two log likelihoods.

We now evaluate the assumption that ε_t^2 follows a standardized GED with shape parameter β. The GED distribution is symmetric and can have fat tails depending on the shape parameter.

In this case, the log-likelihood function takes the following form:

$$\ln L = T \left[\ln \left(\frac{\beta}{\lambda} \right) - \left(1 + \frac{1}{\beta} \right) \ln(2) - \ln \left\{ \Gamma \left(\frac{1}{\beta} \right) \right\} \right] - \frac{1}{2} \sum_{t=1}^{T} \left| \left(\frac{\varepsilon_t}{\lambda \sqrt{h_t}} \right) \right|^{\beta} - \frac{1}{2} \sum_{t=1}^{T} \ln h_t$$

$$(3.6)$$

with $\lambda = \{\Gamma(1/\beta)/2^{2/\beta}\Gamma(3/\beta)\}^{0.5}$, Γ being the gamma function, and $\beta > 0$.

The shape parameter β controls the thickness of the tails, with $\beta = 2$ identifying a normal distribution. When $0 < \beta < 2$, we get distributions whose tails are fatter than the normal case would require (leptokurtic distribution). When $\beta = 1$, we obtain the special case of a Laplace distribution. To fit an ARCH model under the assumption that innovations follow a GED distribution, we must specify the `distribution(ged)` option.

```
. arch return, arch(1) distribution(ged) nolog
```

ARCH family regression

Sample: 04jan1950 - 31dec2013   Number of obs   =   16,102
Distribution: GED   Wald chi2(.)   =   .
Log likelihood = 54045.71   Prob > chi2   =   .

return	Coef.	OPG Std. Err.	z	P>|z|	[95% Conf. Interval]	
return						
_cons	.000533	.0000512	10.41	0.000	.0004326	.0006333
ARCH						
arch						
L1.	.2982344	.0179745	16.59	0.000	.263005	.3334639
_cons	.0000628	1.21e-06	51.82	0.000	.0000604	.0000652
/lnshape	.0502283	.0108383	4.63	0.000	.0289856	.0714711
shape	1.051511	.0113966			1.02941	1.074087

```
. estimates store ARCH1GED
```

In the output reported above, we can see that the shape parameter β for our returns is equal to 1.05, indicating once again that our empirical distribution is far from being normal.

We now obtain the information criteria for the ARCH(1) model with GED innovations to compare it with the ARCH(1) model fit under the normal and the Student's t distributions.

```
. estimates stats ARCH ARCH1T ARCH1GED
```

Akaike's information criterion and Bayesian information criterion

Model	Obs	ll(null)	ll(model)	df	AIC	BIC
ARCH	16,102	.	52723.7	3	-105441.4	-105418.3
ARCH1T	16,102	.	54105.25	4	-108202.5	-108171.7
ARCH1GED	16,102	.	54045.71	4	-108083.4	-108052.7

Note: N=Obs used in calculating BIC; see [R] BIC note.

Under the GED assumption, the AIC is equal to $-108{,}083.4$ and the BIC is equal to $-108{,}052.7$. These are a bit larger than the case where we assume the Student's t distribution. Therefore, we can conclude that the distribution best describing the properties of the S&P daily returns is the Student's t with 3.63 degrees of freedom.

3.2.2   Additional options

The `arch` command has additional options that we can specify:

- `ar(`*numlist*`)` specifies the number of autoregressive terms and `ma(`*numlist*`)` specifies the number of moving-average terms. Alternatively, we can use `arima(`$\#_p$, $\#_d$, $\#_q$`)` to fit an ARIMA(p, d, q) model for the mean equation.

- `het(`*varlist*`)` includes *varlist* in the specification of the conditional variance.

- `vce(`*vcetype*`)` specifies how to compute standard errors. *vcetype* may be one of `opg`, `robust`, or `oim`.

- *maximize_options* control the maximization process.

We now discuss and illustrate these options.

ARIMA

When presenting ARMA models in chapter 2, we analyzed the case of daily returns of the S&P 500, and we provided evidence that the best model in that case was the ARMA(2,0). We now proceed by augmenting the ARCH(1) model by specifying an ARMA(2,0) for the mean equation, always under the assumption that innovations follow a Student's t distribution.

```
. arch return, arch(1) distribution(t) ar(1/2) nolog
ARCH family regression -- AR disturbances
Sample: 04jan1950 - 31dec2013                    Number of obs   =     16,102
Distribution: t                                  Wald chi2(2)    =     170.33
Log likelihood =    54171.9                      Prob > chi2     =     0.0000
```

return	Coef.	OPG Std. Err.	z	P>|z|	[95% Conf. Interval]	
return						
_cons	.0005188	.0000603	8.60	0.000	.0004006	.000637
ARMA						
ar						
L1.	.0938934	.0081124	11.57	0.000	.0779933	.1097934
L2.	-.0413144	.0060046	-6.88	0.000	-.0530831	-.0295457
ARCH						
arch						
L1.	.3416621	.0226364	15.09	0.000	.2972956	.3860286
_cons	.0000674	2.21e-06	30.50	0.000	.0000631	.0000718
/lndfm2	.4913474	.0718089	6.84	0.000	.3506046	.6320903
df	3.634517	.1173729			3.419926	3.881539

The coefficients associated with the mean equation are both statistically significant although quite different from the coefficients that we obtained when fitting a pure
ARMA model in chapter 2: AR(1) was equal to 0.0297 and AR(2) to -0.0429. When
adopting an ARMA–ARCH model, the estimation algorithm carries out the estimation
of the parameters governing the mean and variance equations simultaneously so that
ARMA coefficients can differ even substantially from the case when we consider a simple
mean model with constant variance for the innovations. The ARCH coefficient is instead
quite close to the simple ARCH(1) model with Student's t distribution that we fit in the
previous section. We now evaluate the statistical relevance of adding the ARMA(2,0)
component to the full model.

```
. estat ic
```

Akaike's information criterion and Bayesian information criterion

Model	Obs	ll(null)	ll(model)	df	AIC	BIC
.	16,102	.	54171.9	6	-108331.8	-108285.7

Note: N=Obs used in calculating BIC; see [R] BIC note.

The AIC and the BIC are, respectively, equal to $-108,331.8$ and $-108,285.7$ for the
ARMA(2,0)–ARCH(1). These values are much lower than the ARCH(1) model, leading us
to prefer the ARMA(2,0)–ARCH(1) over the ARCH(1) model.

The het() option

The `het()` option allows us to add some additional explanatory variables to the variance
equation. The variables specified enter the variance model collectively as multiplicative
heteroskedasticity. To illustrate, let us assume that the conditional variance depends on
variables x_t and w_t and that we want to add them to an ARCH(1) model. The conditional
variance equation is

$$h_t = \exp\left(\lambda_0 + \lambda_1 x_t + \lambda_2 w_t\right) + \alpha \varepsilon_{t-1}^2 \tag{3.7}$$

By using the `het()` option with the `arch` command, we could evaluate the presence of
the "weekend effect" in the S&P 500 returns. The weekend effect relies on the empirical
evidence that volatility on Friday is usually lower than volatility on Monday. There
are two reasons that we can ascribe to this effect. First, companies tend to release bad
news on Fridays, after the market closes, so this news will impact on Monday. Second,
short sellers can contribute as well in generating the weekend effect. With the aim of
reducing their risk exposure, they buy the stocks on Fridays, closing their position, and
reopen new short positions on Mondays.

To test for the presence of the weekend effect in our data, we first have to generate
two dummy variables taking a value of 1 for the specific day of the week and a value
of 0 otherwise. We therefore proceed as follows:

```
. generate dayofweek = dow(newdate)

. generate monday = 0

. replace monday = 1 if dayofweek == 1
(2,300 real changes made)

. generate friday = 0

. replace friday = 1 if dayofweek == 5
(2,301 real changes made)

. arch return, arch(1) distribution(t) ar(1/2) het(monday friday) nolog
```

ARCH family regression -- ARMA disturbances and mult. heteroskedasticity

Sample: 04jan1950 - 31dec2013	Number of obs =	16,102
Distribution: t	Wald chi2(2) =	170.26
Log likelihood = 54171.93	Prob > chi2 =	0.0000

return	Coef.	OPG Std. Err.	z	P>|z|	[95% Conf. Interval]	
return						
_cons	.0005188	.0000603	8.60	0.000	.0004006	.000637
ARMA						
ar						
L1.	.0938953	.0081138	11.57	0.000	.0779925	.109798
L2.	-.0412949	.0060053	-6.88	0.000	-.0530651	-.0295246
HET						
monday	.0110139	.0506446	0.22	0.828	-.0882477	.1102755
friday	-.0019901	.0515975	-0.04	0.969	-.1031193	.0991391
_cons	-9.605882	.0343943	-279.29	0.000	-9.673293	-9.53847
ARCH						
arch						
L1.	.3416242	.0226325	15.09	0.000	.2972652	.3859832
/lndfm2	.4915417	.071839	6.84	0.000	.3507397	.6323436
df	3.634835	.117445			3.420118	3.882016

First, we create the variable dayofweek, taking the value of 1 for Monday, 2 for Tuesday, ..., 5 for Friday by means of the variable dow. We then generate the two dummy variables, monday and friday, which after being initialized to 0, are set equal to 1 on the basis of the values taken by dayofweek: when dayofweek is equal to 1, then Monday is set to 1, while when dayofweek is equal to 5, then Friday is set to 1.

Results confirm the presence of the weekend effect. The coefficient associated with Monday is positive and slightly statistically significant, therefore indicating that stock returns are more volatile on Mondays than other days, namely, Tuesday, Wednesday, and Thursday. The negative and slightly statistically significant coefficient associated with Fridays confirms that on those days the S&P 500 is less volatile.

This last model is an AR(2)–ARCH(1) with Student's t innovations and augmented by Monday and Friday dummy variables. We can now compare the goodness of fit of this model with the previously fit model with no dummy variables:

```
. estat ic
Akaike's information criterion and Bayesian information criterion
```

Model	Obs	ll(null)	ll(model)	df	AIC	BIC
.	16,102	.	54171.93	8	-108327.9	-108266.4

Note: N=Obs used in calculating BIC; see [R] BIC note.

The AIC for the model augmented by the two dummy variables is equal to −108,327.9, while it was equal to −108,331.8 without the two dummy variables. We get a similar indication when considering the BIC, which is equal to −108,266.4 and −108,285.7, respectively. These results indicate that, although the likelihood for this second model slightly improves, the two dummy variables `monday` and `friday` do not justify the loss of two degrees of freedom.

The maximize_options options

ARCH models are fit by maximum likelihood, with the likelihood functions being, for instance, those specified in (3.4), (3.5), and (3.6), according to the distribution assumption for innovations. When working with ARCH models, we speak about conditional likelihood, because the likelihood depends on a set of priming values assumed for the squared innovations ε_t^2 and variances h_t, which are not part of the estimation sample.

We briefly summarize several ways by which a maximization procedure can be carried out. For further information, see [TS] **arch**.

- `difficult` uses a different stepping algorithm in nonconcave regions. We advise using the `difficult` option when the last iteration is not concave or when we receive during the maximization procedure the warning "not concave", indicating that the standard stepping maximization algorithm is not working well.

- `technique(`*algorithm_spec*`)` specifies the maximization technique: `nr` (Newton–Raphson algorithm, the default), `bhhh` (Berndt–Hall–Hall–Hausman algorithm), `dfp` (Davidon–Fletcher–Powell algorithm), or `bfgs` (Broyden–Fletcher–Goldfarb–Shanno algorithm). We can even specify a sequence of algorithms to be used during the maximization procedure with the number of iterations for each algorithm. For instance, by indicating `technique(bhhh 10 nr 1000)`, we are requesting that Stata perform 10 iterations with the BHHH algorithm followed by 1,000 iterations with the NR algorithm, and after that switching back to BHHH for 10 iterations, and so on.

- `from(`*init_specs*`)` specifies the initial values of the coefficients. ARCH models may be sensitive to initial values and may have coefficient values that correspond to local maximums. The default starting values are obtained with a series of regressions, producing results that, on the basis of asymptotic theory, are consistent for the β and ARMA parameters and are generally reasonable for the rest. Nevertheless, these values may not always be feasible in that the likelihood function cannot be evaluated at the initial values the `arch` command chooses at first. In this case, the estimation command restarts with ARCH and ARMA parameters initialized to 0. It is possible that even these values will be infeasible and that we will have to supply initial values. We can supply our initial values as a matrix, a list of values, or coefficient name–value pairs.

It is not always easy to fit ARCH models, and the difficulties experienced by the maximization algorithm are shown in the log. For instance, we might receive the message "switching optimization to . . . ", indicating that the maximization algorithm cannot find a solution and therefore Stata chooses to use an alternative algorithm. "Backed up" is a typical message obtained from BFGS stepping because the BFGS Hessian is often overoptimistic. When we receive this message, we must check whether the log likelihood is improving because it could be a signal of a problem. In that case, it is advisable to reset the gradient tolerance to a larger value, although that can lead to a solution that is not the global maximum.

Quite often, we will receive the message "BFGS stepping has contracted, resetting BFGS Hessian", which may indicate a situation in which the algorithm is stuck and is experiencing difficulties in improving the log-likelihood value. In technical language, it indicates that the algorithm resets its Hessian and takes a steepest-descent step. We should take care when we receive this message repeatedly because it can indicate that the algorithm will not find a solution. In this case, it can be meaningful to stop the iterative procedure and set the maximum number of iterations to that point where the algorithm stopped using the command; the algorithm will then stop at that point, and we will be in a position to investigate why the maximization cannot carry on.

3.2.3 Postestimation

A simple ARCH(1) model will not be able to capture the time structure of the time series in analysis. To check whether an ARCH(1) model is correctly specified and therefore all the data characteristics have been modeled, we can start by evaluating the properties of its standardized residuals. When we introduced ARCH models in (3.1), we required standardized residuals $\varepsilon_t = \eta_t \sqrt{h_t}$ to follow a white-noise process. This condition holds true if innovations satisfy the following conditions:

- do not show any heteroskedasticity, because volatility clusters must have been completely modeled by GARCH;

- are serial uncorrelated; and

- are distributed like the prespecified distribution (for instance, the Gaussian distribution).

We are now going to evaluate whether standardized residuals meet these conditions. Before starting any analysis, we need to get standardized residuals. We fit an ARCH(1) model under the normal assumption, and then we get residuals and variance by using the `predict` command.

```
arch return, arch(1) nolog
predict residual, residuals
predict variance, variance
generate stdresidual = residual/variance^0.5
```

We obtain residuals by specifying the `residuals` option with the `predict` command, and then we obtain the variance by using the `variance` option. Finally, we compute the standardized residuals as the ratio between `residual` and the square root of `variance`.

Once we compute the standardized residuals, we can check whether they satisfy the three conditions stated above. We start by evaluating whether the standardized residuals still show some heteroskedasticity. We run a regression with just the constant as an explanatory variable, and then, by using the `estat archlm` command, we conduct the LM–ARCH test with $(1, 5, 10)$ lags.

```
. regress stdresidual
```

Source	SS	df	MS		Number of obs	=	16,102
					F(0, 16101)	=	0.00
Model	0	0	.		Prob > F	=	.
Residual	16096.9847	16,101	.999750615		R-squared	=	0.0000
					Adj R-squared	=	0.0000
Total	16096.9847	16,101	.999750615		Root MSE	=	.99988

stdresidual	Coef.	Std. Err.	t	P>|t|	[95% Conf. Interval]	
_cons	-.0176403	.0078796	-2.24	0.025	-.0330853	-.0021954

```
. estat archlm, lags(1,5,10)
LM test for autoregressive conditional heteroskedasticity (ARCH)
```

lags(p)	chi2	df	Prob > chi2
1	5.440	1	0.0197
5	1311.531	5	0.0000
10	1744.360	10	0.0000

H0: no ARCH effects   *vs.* H1: ARCH(p) disturbance

The results lead us to reject the null hypothesis of no ARCH effect in the standardized residuals, indicating that a simple ARCH(1) model cannot completely remove all the heteroskedasticity characterizing our data. This finding suggests we add extra lags in the ARCH model to model a richer time dependence of our time series.

The second assumption that standardized residuals must fulfill is the absence of any autocorrelation structure. To check this, we obtain the autocorrelogram of standardized residuals by using the `ac` command:

```
. ac stdresidual
```

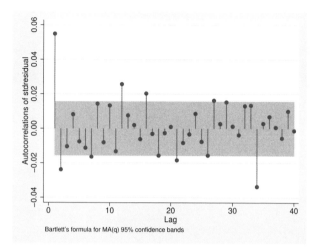

Figure 3.2. Autocorrelogram for standardized residuals of ARCH(1) model

In figure 3.2, we can see that autocorrelations at some lags are statistically different from 0, indicating that standardized residuals still show some autocorrelation. Therefore, the second condition for standardized residuals is also not satisfied.

The last condition to check is whether the standardized residuals follow the distribution under which the estimation is carried out. In chapter 1, we presented several ways to test for normality. Here we proceed by comparing the distribution of standardized residuals with the theoretical one, which in our case is the normal distribution, by using the `histogram` command.

```
. histogram stdresidual, normal normopts(lcolor(gs13)) kdensity
(bin=42, start=-11.221331, width=.51172302)
```

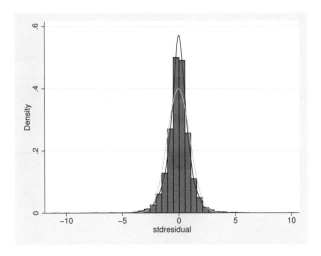

Figure 3.3. Distribution for standardized residuals of ARCH(1) model

In figure 3.3, the light gray line indicates the normal distribution, and the dark gray line identifies the empirical one. From a visual inspection, we can see that the two distributions are quite different, with the empirical one being more peaked than the normal and suggesting the need to switch to a fat-tailed distribution. The normal assumption is therefore rejected.

After evaluating the properties of standardized residuals, we conclude that none of them is met, with the consequence that our ARCH(1) model for the conditional variance is not correctly specified. We therefore need to move to alternative and more sophisticated models.

A final important point when assessing whether the model is correctly specified is to evaluate the presence of outliers because they can dramatically influence the tests for heteroskedasticity. To test for the presence of outliers, we can use the following procedure:

1. Fit the GARCH model and obtain the quantity $\widehat{v}_t = r_t^2 - \widehat{h}_t$, which measures how big outliers are.

2. Compare the quantity obtained in step 1 with a predefined threshold.

3. Replace outliers with a correct estimate.

4. Fit the model again for the series on the corrected time series.

3.3   ARCH(p)

An ARCH(1) model generally provides a poor fit to financial time series. A natural extension of this model is to add additional p lags to obtain an ARCH(p) model:

$$h_t = \omega + \sum_{i=1}^{p} \alpha_i \varepsilon_{t-i}^2 \qquad (3.8)$$

The model in (3.8) satisfies the conditions $E(h_t) = \sigma^2$ and $\sigma^2 = \omega/1 - \sum_{i=1}^{p} \alpha_i$, where $\omega > 0$, $\alpha_i \geq 0$ for $i = 1, \ldots, p$, and $\sum_{i=1}^{p} \alpha_i < 1$.

We now proceed by adding, say, 9 extra lags to our initial ARCH(1) model for the S&P daily returns.

```
. arch return, arch(1/10) nolog

ARCH family regression

Sample: 04jan1950 - 31dec2013                    Number of obs   =     16,102
Distribution: Gaussian                           Wald chi2(.)    =          .
Log likelihood =    54654.3                       Prob > chi2     =          .
```

return	Coef.	OPG Std. Err.	z	P>|z|	[95% Conf. Interval]	
return						
_cons	.000531	.0000545	9.74	0.000	.0004242	.0006379
ARCH						
arch						
L1.	.1161806	.0047579	24.42	0.000	.1068554	.1255058
L2.	.0942936	.0065369	14.42	0.000	.0814816	.1071057
L3.	.0808583	.0071961	11.24	0.000	.0667543	.0949623
L4.	.0840873	.0069045	12.18	0.000	.0705547	.09762
L5.	.0842436	.0082924	10.16	0.000	.0679909	.1004963
L6.	.0652805	.0068085	9.59	0.000	.051936	.0786249
L7.	.0711151	.0067394	10.55	0.000	.0579062	.0843241
L8.	.0996053	.0061157	16.29	0.000	.0876187	.1115919
L9.	.0980128	.0055522	17.65	0.000	.0871307	.1088949
L10.	.0600367	.0070771	8.48	0.000	.0461658	.0739075
_cons	.0000165	5.38e-07	30.57	0.000	.0000154	.0000175

All the coefficients are statistically significant, suggesting that the ARCH(10) model should outperform the ARCH(1), at least from a goodness-of-fit point of view. We can compare the two models in a more formal way, adopting the information criteria:

```
. estat ic
Akaike's information criterion and Bayesian information criterion
```

Model	Obs	ll(null)	ll(model)	df	AIC	BIC
.	16,102	.	54654.3	12	-109284.6	-109192.4

Note: N=Obs used in calculating BIC; see [R] BIC note.

The information criteria clearly indicate that we must prefer the ARCH(10) model over the ARCH(1) model: the ARCH(10) AIC equals $-109{,}284.6$ versus $-105{,}441.4$ for the ARCH(1), and the ARCH(10) BIC equals $-109{,}192.4$ versus $-105{,}418.3$ for the ARCH(1).

Now that we have verified that the ARCH(10) model outperforms the ARCH(1), we have to check whether this model is correctly specified, starting by evaluating whether the standardized residuals still show any ARCH effect. We compute standardized residuals as shown above, run the regression on standardized residuals with just the constant as an explanatory variable, and then conduct the LM–ARCH test that we report below.

```
. predict residual, residuals
(1 missing value generated)
. predict variance, variance
. generate stdresidual = residual/variance^0.5
(1 missing value generated)
. regress stdresidual
```

Source	SS	df	MS			
				Number of obs	=	16,102
				F(0, 16101)	=	0.00
Model	0	0	.	Prob > F	=	.
Residual	16085.1437	16,101	.999015196	R-squared	=	0.0000
				Adj R-squared	=	0.0000
Total	16085.1437	16,101	.999015196	Root MSE	=	.99951

stdresidual	Coef.	Std. Err.	t	P>|t|	[95% Conf. Interval]	
_cons	-.0322735	.0078767	-4.10	0.000	-.0477128	-.0168342

```
. estat archlm, lags(1,5,10)
LM test for autoregressive conditional heteroskedasticity (ARCH)
```

lags(*p*)	chi2	df	Prob > chi2
1	0.081	1	0.7759
5	4.620	5	0.4640
10	29.951	10	0.0009

```
        H0: no ARCH effects      vs.  H1: ARCH(p) disturbance
. drop residual variance stdresidual
```

The null hypothesis of no ARCH effect is not rejected at 1 and 5 lags, but it is still rejected when taking into account 10 lags, indicating that an ARCH(10) model cannot completely describe our data.

The results show that the ARCH(10) model is not correctly specified, but this is not the only reason we are not going to select this model. In fact, an ARCH(10) model is an overparameterized one that cannot be usefully adopted to make predictions; it is too data driven, and parameter estimates can suffer from estimation problems. Moreover, despite the high number of parameters added into the model, there is still some heteroskedasticity left in the data that suggests we should add some extra parameters, leading the number of lags to be far beyond 10.

The issue with ARCH models having a high number of parameters can be easily addressed by switching to GARCH models.

3.4 GARCH models

3.4.1 GARCH(p,q)

Bollerslev (1986) introduced the generalized autoregressive conditional heteroskedasticity (GARCH) model to deal with the dimensionality problem affecting ARCH(p) models. This issue was addressed by using lags of conditional variance as regressors. We can represent a GARCH(p, q) model as

$$h_t = \omega + \sum_{i=1}^{p} \alpha_i \varepsilon_{t-i}^2 + \sum_{j=1}^{q} \beta_j h_{t-j}$$

where $\omega > 0$, $\alpha_i \geq 0$ with $i = 1, \ldots, p$, and $\beta_j \geq 0$ with $j = 1, \ldots, q$. Here p indicates the number of lags for the squared residuals, and q indicates the number of lags for the conditional variance. We can consider the GARCH model to be of the class of ARMA models, where the conditional variance follows an ARMA process with (p, q) terms. The underlying idea of a GARCH model is to ensure its parsimony by adding lags of conditional variance.

Generally, the GARCH(1,1) specification is the most appropriate representation to describe the conditional volatility process. It takes the following form:

$$h_t = \omega + \alpha \varepsilon_{t-1}^2 + \beta h_{t-1} \tag{3.9}$$

where $\alpha + \beta < 1$. The α parameter typically takes a value between 0 and 0.1, and β takes a value between 0.75 and 0.95, according to the persistence of the time series.

We now calculate the expected value of a GARCH(1,1) model.

$$
\begin{aligned}
E\left(\varepsilon_t\right) &= E\left(\eta \sqrt{h_t}\right) \\
E\left(\varepsilon_t\right) &= E\left(\eta_t \sqrt{\omega + \alpha \varepsilon_{t-1}^2 + \beta h_{t-1}}\right) \\
E\left(\varepsilon_t\right) &= E\left(\eta_t\right) E\left(\sqrt{\omega + \alpha \varepsilon_{t-1}^2 + \beta h_{t-1}}\right) \\
E\left(\varepsilon_t\right) &= 0
\end{aligned}
$$

We can obtain the unconditional or long-run variance as

$$\sigma^2 = \text{Var}(\varepsilon_t) = E\left\{\varepsilon_t - E\left(\varepsilon_t\right)\right\}^2$$
$$\sigma^2 = E\left(\varepsilon_t\right)^2$$
$$\sigma^2 = E\left\{\eta_t^2\left(\omega + \alpha\varepsilon_{t-1}^2 + \beta h_{t-1}\right)\right\}$$
$$\sigma^2 = E\left(\eta_t^2\omega\right) + E\left(\eta_t^2\alpha\varepsilon_{t-1}^2\right) + E\left(\eta_t^2\beta h_{t-1}\right)$$
$$\sigma^2 = \omega + \alpha E\left(\varepsilon_{t-1}^2\right) + \beta E\left(h_{t-1}\right)$$
$$\sigma^2 = \omega + \alpha\sigma^2 + \beta\sigma^2$$

allowing us to rewrite the unconditional variance as

$$\sigma^2 = \omega + \alpha\sigma^2 + \beta\sigma^2$$
$$\sigma^2 = \frac{\omega}{1-\alpha-\beta}$$

which is well defined if $\omega > 0$ and $0 \leq \alpha + \beta < 1$.

We can rewrite a GARCH(1,1) model as an infinite ARCH process. To illustrate this point, let us define

$$h_t = \omega + \alpha\varepsilon_{t-1}^2 + \beta h_{t-1}$$
$$h_{t-1} = \omega + \alpha\varepsilon_{t-2}^2 + \beta h_{t-2}$$

Substituting the expression for h_{t-1} in the equation for h_t, we get

$$h_t = \omega + \alpha\varepsilon_{t-1}^2 + \beta\left(\omega + \alpha\varepsilon_{t-2}^2 + \beta h_{t-2}\right)$$
$$h_t = \omega\left(1 + \beta\right) + \alpha\varepsilon_{t-1}^2 + \beta\alpha\varepsilon_{t-2}^2 + \beta^2 h_{t-2}$$

Proceeding in this way, we obtain

$$h_t = \sum_i^\infty \beta^i\omega + \sum_i^\infty \beta^{i-1}\varepsilon_{t-i}^2$$

The conditional variance at time t is a function of a constant ω and an infinite number of lagged squared residuals ε^2.

We now come back to the empirical application and fit a GARCH(1,1) model by specifying the `garch()` option with the `arch` command, as follows:

```
. arch return, arch(1) garch(1) nolog
ARCH family regression
Sample: 04jan1950 - 31dec2013                    Number of obs   =      16,102
Distribution: Gaussian                           Wald chi2(.)    =           .
Log likelihood =  54784.48                       Prob > chi2     =           .
```

return	Coef.	OPG Std. Err.	z	P>|z|	[95% Conf. Interval]	
return						
_cons	.0004798	.0000537	8.93	0.000	.0003745	.000585
ARCH						
arch L1.	.0817023	.0016294	50.14	0.000	.0785086	.0848959
garch L1.	.9118854	.0021948	415.48	0.000	.9075837	.9161871
_cons	8.08e-07	6.70e-08	12.06	0.000	6.77e-07	9.39e-07

Both ARCH and GARCH coefficients are statistically different from 0, with their sum being smaller but close to 1 and therefore satisfying the stationarity condition of GARCH models. We can even compare the GARCH(1,1) model with both the ARCH(1) and the ARCH(10) models by means of their information criteria.

```
. estat ic
Akaike´s information criterion and Bayesian information criterion
```

Model	Obs	ll(null)	ll(model)	df	AIC	BIC
.	16,102	.	54784.48	4	-109561	-109530.2

Note: N=Obs used in calculating BIC; see [R] BIC note.

The GARCH model outperforms both ARCH models. By comparing the log likelihoods for ARCH(10) and GARCH(1,1), we can see that they are quite close, being equal to 54,654.3 and 54,784.5, respectively. The ARCH(10) model has 11 parameters, while the GARCH model has just 3 parameters, leading the information criteria to clearly prefer the GARCH(1,1) model.

Now we check whether the standardized residuals satisfy the requirement of no longer exhibiting heteroskedasticity by applying the LM–ARCH test.

```
. predict residual, residuals
(1 missing value generated)
. predict variance, variance
. generate stdresidual = residual/variance^0.5
(1 missing value generated)
```

```
. regress stdresidual
```

(output omitted)

```
. estat archlm, lags(1,5,10)
LM test for autoregressive conditional heteroskedasticity (ARCH)
```

lags(p)	chi2	df	Prob > chi2
1	5.763	1	0.0164
5	6.234	5	0.2841
10	11.661	10	0.3084

```
         H0: no ARCH effects     vs.  H1: ARCH(p) disturbance
. drop residual variance stdresidual
```

We do not reject the null hypothesis of absence of heteroskedasticity at 5 and 10 lags, but we still reject the null hypothesis at the first lag.

To improve the goodness of fit of our model, we add an additional lag for the ARCH term and one for the GARCH term. We therefore proceed by fitting a GARCH(2,2) model, which is a model with two lags for the ARCH part and two terms for the GARCH part. We fit this model by using the `arch` command with the options `arch()` and `garch()`, both of them taking 1/2 as the argument, indicating that we want to add two lags. Once the GARCH(2,2) model is fit, we obtain standardized residuals and conduct the LM test for absence of autoregressive conditional heteroskedasticity.

```
. arch return, arch(1/2) garch(1/2) nolog
ARCH family regression
Sample: 04jan1950 - 31dec2013                Number of obs   =    16,102
Distribution: Gaussian                       Wald chi2(.)    =         .
Log likelihood =  54816.78                   Prob > chi2     =         .
```

return	Coef.	OPG Std. Err.	z	P>|z|	[95% Conf. Interval]	
return						
_cons	.0004986	.0000554	9.00	0.000	.0003901	.0006072
ARCH						
arch						
L1.	.096543	.0022919	42.12	0.000	.092051	.101035
L2.	-.0944451	.0022137	-42.66	0.000	-.0987838	-.0901064
garch						
L1.	1.845729	.0080074	230.50	0.000	1.830035	1.861424
L2.	-.8479704	.0077176	-109.87	0.000	-.8630967	-.8328441
_cons	1.56e-08	3.28e-09	4.76	0.000	9.18e-09	2.20e-08

```
. predict residual, residuals
(1 missing value generated)

. predict variance, variance

. generate stdresidual = residual/variance^0.5
(1 missing value generated)
```

```
. regress stdresidual
  (output omitted)
. estat archlm, lags(1,5,10)
LM test for autoregressive conditional heteroskedasticity (ARCH)
```

lags(p)	chi2	df	Prob > chi2
1	2.521	1	0.1123
5	3.242	5	0.6627
10	9.538	10	0.4819

H0: no ARCH effects   *vs.* H1: ARCH(p) disturbance

The two additional lags, the ARCH(2) and the GARCH(2) are both statistically significant. Moreover, the standardized residuals obtained from a GARCH(2,2) model do not display any heteroskedasticity because we do not reject the null hypothesis of absence of heteroskedasticity at all the lags considered. We can therefore conclude that the GARCH(2,2) model is correctly specified. The final step is to check whether the GARCH(2,2) model outperforms the GARCH(1,1) by checking the information criteria.

```
. quietly arch return, arch(1/2) garch(1/2) nolog
. estat ic
Akaike's information criterion and Bayesian information criterion
```

Model	Obs	ll(null)	ll(model)	df	AIC	BIC
.	16,102	.	54816.78	6	-109621.6	-109575.4

Note: N=Obs used in calculating BIC; see [R] BIC note.

The AIC for the GARCH(2,2) model is $-109{,}621.6$ versus $-109{,}561$ for the GARCH(1,1). The BIC is equal to $-109{,}575.4$ and $-109{,}530.2$, respectively. Both indicate that we should prefer the GARCH(2,2) model over the GARCH(1,1).

We now evaluate whether the standardized residuals of the GARCH(2,2) model display any autocorrelation by analyzing their autocorrelogram, shown in figure 3.4:

```
. ac stdresidual
```

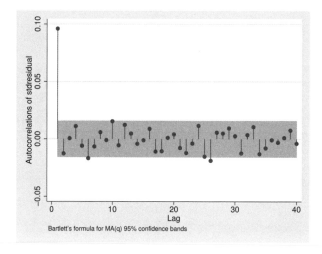

Figure 3.4. Autocorrelogram for residuals from GARCH(2,2)

We can see that all the autocorrelations except the first one lie inside the gray shaded area or arbitrarily close to it, indicating that we have evidence of autocorrelation just at the first lag.

Finally, we compare the distribution of standardized residuals with the theoretical one (which in our case is the normal distribution) by using the `histogram` command:

```
. histogram stdresidual, normal normopts(lcolor(gs13)) kdensity
(bin=42, start=-11.127096, width=.41976547)
```

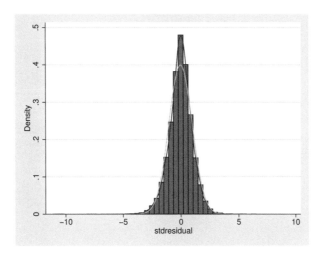

Figure 3.5. Histogram for residuals from GARCH(2,2)

From figure 3.5, we see that the empirical distribution of standardized residuals is still more peaked than the normal distribution, although here, the similarity between the theoretical and the empirical distribution is much closer than in the case of the ARCH(1) model, depicted in figure 3.3.

In chapter 2, we showed that the best model for the mean equation of the S&P 500 daily returns is the ARMA(2,0). When introducing the ARCH model, we showed that the distribution best fitting the empirical data is the Student's t. We therefore now fit an AR(2)–GARCH(2,2) model with a Student's t distribution:

```
. drop residual variance stdresidual
. arch return, arch(1/2) garch(1/2) ar(1/2) distribution(t) nolog
ARCH family regression -- AR disturbances
Sample: 04jan1950 - 31dec2013                Number of obs   =    16,102
Distribution: t                              Wald chi2(2)    =    151.83
Log likelihood =  55309.85                   Prob > chi2     =    0.0000
```

return	Coef.	OPG Std. Err.	z	P>|z|	[95% Conf. Interval]	
return						
_cons	.0005569	.0000557	9.99	0.000	.0004476	.0006661
ARMA						
ar						
L1.	.0952509	.0081117	11.74	0.000	.0793522	.1111496
L2.	-.0388044	.0079824	-4.86	0.000	-.0544497	-.0231592
ARCH						
arch						
L1.	.0771468	.0046131	16.72	0.000	.0681052	.0861883
L2.	.0717009	.0053158	13.49	0.000	.0612821	.0821198
garch						
L1.	-.0604492	.0169706	-3.56	0.000	-.093711	-.0271874
L2.	.9021064	.0152604	59.11	0.000	.8721964	.9320163
_cons	1.20e-06	1.73e-07	6.96	0.000	8.64e-07	1.54e-06
/lndfm2	1.568189	.0638816	24.55	0.000	1.442983	1.693395
df	6.797951	.3065009			6.233306	7.437909

We can see that all the parameters are statistically significant, including the number of degrees of freedom characterizing the Student's t distribution.

The information criteria for this model are as follows:

```
. estat ic
Akaike's information criterion and Bayesian information criterion
```

Model	Obs	ll(null)	ll(model)	df	AIC	BIC
.	16,102	.	55309.85	9	-110601.7	-110532.5

Note: N=Obs used in calculating BIC; see [R] BIC note.

Both information criteria are far lower than those obtained for the GARCH(2,2) model under the assumption of Gaussianity and with no model for the mean part.

An interesting relationship exists between the resampling frequency and the β parameter accounting for persistence in the GARCH model. To illustrate this point, we now fit a GARCH(1,1) model on monthly returns using a resampled time series of our daily S&P 500 returns, under the Gaussianity assumption and with no mean equation.

```
. use http://www.stata-press.com/data/feus/spmonthly, clear

. arch return, arch(1) garch(1) nolog

ARCH family regression

Sample: 2 - 769                               Number of obs   =        768
Distribution: Gaussian                        Wald chi2(.)    =         .
Log likelihood =  1372.229                    Prob > chi2     =         .
```

return	Coef.	OPG Std. Err.	z	P>|z|	[95% Conf. Interval]	
return						
_cons	.0065792	.001475	4.46	0.000	.0036883	.0094702
ARCH						
arch						
L1.	.1130881	.0245265	4.61	0.000	.0650171	.1611591
garch						
L1.	.8405205	.0279828	30.04	0.000	.7856753	.8953657
_cons	.000092	.0000315	2.92	0.004	.0000303	.0001538

The β parameter is equal to 0.84, while it equaled 0.91 when we fit the model on daily data. Therefore, as expected, we confirm that the variance process is more persistent when measured on higher frequency data.

A peculiar case of GARCH models is the integrated GARCH (IGARCH) model, which is characterized by the presence of a unit root in the autoregressive dynamic of squared residuals, corresponding to setting $\alpha + \beta = 1$ in (3.9). The IGARCH(1,1) model takes the following form:

$$h_t = \omega + \alpha \varepsilon_{t-1}^2 + (1 - \alpha) h_{t-1}$$

Given that the IGARCH model is nonstationary, this process is useful when the conditional variance is highly serially correlated (long-memory process), for instance, when working with intraday data.

An example of the IGARCH model is the risk metrics model. In this case, the values of the ARCH and GARCH parameters are fixed: $\alpha = (1 - \lambda)$ and $\beta = \lambda$.

$$h_t = (1 - \lambda) \varepsilon_{t-1}^2 + \lambda h_{t-1}$$

where $\omega = 0$, $\lambda = 0.94$ for daily data, and $\lambda = 0.97$ for weekly data.

3.4.2 GARCH in mean

The conditional variance can even enter the equation for the conditional mean. In that case, we have a GARCH-in-mean (GARCH-M) model. The GARCH-M model was proposed to allow us to account for the widely studied relationship between risk and return: as the volatility of an asset raises, so does the expected risk premium. We can represent the GARCH-M as follows:

$$r_t = \omega + \beta x_t + \theta h_t + \varepsilon_t \tag{3.10}$$

where h_t follows a GARCH process, θ is the risk aversion parameter, and x_t is the vector of exogenous variables at time t.

Instead of the linear form in (3.10), we can insert the conditional variance h_t in the equation for the conditional mean by adopting a nonlinear function $g(\cdot)$:

$$r_t = \omega + \beta x_t + \theta_0 g(h_t) + \theta_1 g(h_{t-1}) + \theta_2 g(h_{t-2}) + \cdots + \varepsilon_t$$

We can fit an ARCH-in-mean (ARCH-M) model by specifying the **archm** option in the usual **arch** command. Then, the **archmlags**(*numlist*) option specifies the number of lags for the conditional variance that we want to add in the conditional mean equation. For instance, by specifying **archmlags**(0), we add just the contemporaneous conditional variance h_t; by specifying **archmlags**(1), we are adding the once-lagged variance h_{t-1}.

```
. use http://www.stata-press.com/data/feus/spdaily

. tsset newdate
        time variable:  newdate, 03jan1950 to 31dec2013
                delta:  1 day

. arch return, arch(1) garch(1) archm archmlags(1) nolog

ARCH family regression

Sample: 04jan1950 - 31dec2013                    Number of obs   =      16,102
Distribution: Gaussian                           Wald chi2(2)    =       11.95
Log likelihood =    54792.1                       Prob > chi2     =      0.0025
```

		OPG				
return	Coef.	Std. Err.	z	P>|z|	[95% Conf. Interval]	
return						
_cons	.0003063	.0000786	3.90	0.000	.0001522	.0004605
ARCHM						
sigma2						
--.	17.18942	6.664707	2.58	0.010	4.126831	30.252
L1.	-13.97885	6.509351	-2.15	0.032	-26.73695	-1.220761
ARCH						
arch						
L1.	.081337	.0016464	49.40	0.000	.07811	.0845639
garch						
L1.	.9122545	.0022107	412.65	0.000	.9079216	.9165874
_cons	8.03e-07	6.76e-08	11.89	0.000	6.71e-07	9.35e-07

In the output reported above, we can see the two extra parameters for the ARCH-M part as well as the archmlags(1) option. These two parameters correspond to coefficients in (3.10), loading h_t and h_{t-1}, respectively, and they are both statistically significant.

3.4.3   Forecasting

On the basis of a GARCH(1,1), we can obtain a volatility forecast at time $t+1$ as

$$E\left(h_{t+1}|I_t\right) = E\left(\widehat{\omega} + \widehat{\alpha}\varepsilon_t^2 + \widehat{\beta}h_t|I_t\right)$$
$$E\left(h_{t+1}|I_t\right) = \widehat{\omega} + \widehat{\alpha}\varepsilon_t^2 + \widehat{\beta}h_t$$

where we are exploiting the fact that at time t, given the information set I_t, we know both quantities ε_t and h_t. When moving to forecasts at the next time $t+k$ with $k \geq 2$, it is necessary to distinguish between dynamic and static forecasts.

In the case of dynamic forecast, the informative set remains the same through time, and equal to I_t, which is where the time series stops. For instance, at time $t+2$, the dynamic forecast for the conditional volatility is

$$E(h_{t+2}|I_t) = E\left(\widehat{\omega} + \widehat{\alpha}\varepsilon_{t+1}^2 + \widehat{\beta}h_{t+1}|I_t\right)$$
$$E(h_{t+2}|I_t) = \widehat{\omega} + \widehat{\alpha}\left(\varepsilon_{t+1}^2|I_t\right) + \widehat{\beta}\left(h_{t+1}|I_t\right)$$
$$E(h_{t+2}|I_t) = \widehat{\omega} + \left(\widehat{\alpha} + \widehat{\beta}\right)E\left(h_{t+1}|I_t\right)$$

At a generic lag $t + k$, we can obtain the dynamic forecast as

$$E(h_{t+k}|I_t) = E\left(\widehat{\omega} + \widehat{\alpha}\varepsilon_{t+k-1}^2 + \widehat{\beta}h_{t+k-1}|I_t\right)$$

and after recursive substitutions,

$$E(h_{t+k}|I_t) = \widehat{\omega}\left\{1 + \left(\widehat{\alpha} + \widehat{\beta}\right) + \cdots + \left(\widehat{\alpha} + \widehat{\beta}\right)^{k-2}\right\} + \left(\widehat{\alpha} + \widehat{\beta}\right)^{k-1}E\left(h_{t+1}|I_t\right)$$

where $0 \leq \widehat{\alpha} + \widehat{\beta} < 1$ is exponentially decaying as $k \to \infty$.

For $k \to \infty$, we obtain

$$\lim_{k\to\infty} E\left(h_{t+k}|I_t\right) = \lim_{k\to\infty} \sum_{i=0}^{k-2} \widehat{\omega}\left(\widehat{\alpha} + \widehat{\beta}\right)^i$$
$$\lim_{k\to\infty} E\left(h_{t+k}|I_t\right) = \sum_{i=0}^{\infty} \widehat{\omega}\left(\widehat{\alpha} + \widehat{\beta}\right)^i \qquad (3.11)$$
$$\lim_{k\to\infty} E\left(h_{t+k}|I_t\right) = \frac{\widehat{\omega}}{1-\left(\widehat{\alpha}+\widehat{\beta}\right)}$$

with $0 \leq \widehat{\alpha} + \widehat{\beta} < 1$.

Equation (3.11) indicates that in the long run, the conditional variance converges to the unconditional variance (we obtained the expression for the unconditional variance at the beginning of this section). Therefore, the key takeaway from this result is that any forecasts for the conditional volatility that we can obtain by applying GARCH models are meaningful just in the short run.

Turning now to static forecast, the main difference with respect to the dynamic forecast is that here the informative set changes through time, incorporating recent news. The information set is defined by rolling windows, although parameters are not reestimated as new information becomes available. At time $t + 2$, we can obtain the static forecast as

$$E\left(h_{t+2}|I_{t+1}\right) = \widehat{\omega} + \widehat{\alpha}\varepsilon_{t+1}^2 + \widehat{\beta}h_{t+1|t}$$

while at the generic time horizon k, the static forecast is

$$E\left(h_{t+k}|I_{t+k-1}\right) = \widehat{\omega} + \widehat{\alpha}\varepsilon_{t+k-1}^2 + \widehat{\beta}h_{t+k-1|t+k-2}$$

As illustrated in chapter 2, we can obtain forecasts by using the `predict` command. In particular, after having fit a GARCH model, we can obtain predictions about

- the mean equation, using the `xb` option (the default);
- the conditional variance, using the `variance` option;
- the multiplicative heteroskedasticity component of variance, using the `het` option;
- residuals, using the `residuals` option, about the modeled variable, that could be a transformation of the original variable y;
- residuals in terms of the mean equation, using the `yresiduals` option. In this case, we obtain forecasts on y, regardless of any transformation adopted for y.

To propose an example, we now want to obtain forecasts for the mean equation on the basis of the previously fit ARMA(2,0)–GARCH(2,2) model with a Student's t distribution. We use the following commands:

```
arch return, arch(1/2) garch(1/2) ar(1/2) distribution(t) nolog
predict mean, xb
tsset date
tsline return mean
```

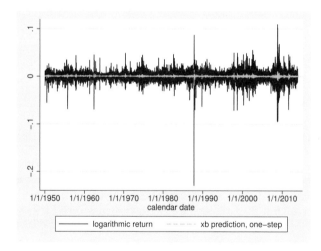

Figure 3.6. Return forecasts from ARMA(2,0)–GARCH(2,2) Student's t model

In the code reported above, we have stored the forecast for the mean equation by using the `xb` option with the `mean` variable. After that, we use the `tsline` command to plot the return time series with the mean forecast in figure 3.6. Unfortunately, the forecast of the level of returns is quite poor, almost corresponding to a straight line throughout the time frame considered.

In addition to the mean equation, we can get the forecast of the conditional variance by specifying the **variance** option (see figure 3.7):

```
predict variance, variance
tsline variance
```

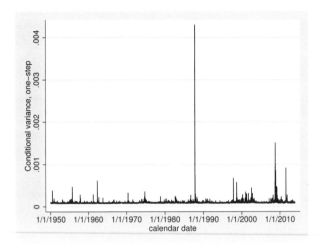

Figure 3.7. Variance forecasts from ARMA(2,0)–GARCH(2,2) Student's t model

3.5   Asymmetric GARCH models

Black (1976) showed that negative news has a greater impact on conditional volatility than does positive news. This is the well-known leverage effect that we can explain as follows. Consider the case of a release of negative news on the future profitability of a firm. Following this news, the firm's stocks will decrease substantially in their value, causing the leverage to increase. A higher level of leverage will lead the perception of the risk associated with a firm to increase with the volatility of its stocks. Therefore, we expect to observe higher volatility when negative news arrives on the market versus when positive news is released.

When the GARCH model is specified, the impact of innovations is symmetric because innovations (regardless of their sign) enter the model with the same coefficient α and leave all the explanatory power to the size of news. To account for the presence of the leverage effect, several authors proposed alternative GARCH models. Before describing asymmetric GARCH models, we first introduce these tests for the presence of the leverage effect in time series: the sign bias test, the negative sign bias test, and the positive sign bias test.

Engle and Ng (1993) propose the sign bias tests relying on the following regression:

$$\widehat{\varepsilon}_t^2 = \beta_0 + \beta_1 \widehat{\gamma}_{t-1} + \xi_t$$

According to the form of $\widehat{\gamma}_{t-1}$, we can obtain alternative tests. By indicating with $I(\cdot)$ the indicator function, we can define

- the sign bias test if $\widehat{\gamma}_{t-1}$ is equal to $I(\widehat{\varepsilon}_{t-1} < 0)$;
- the negative sign bias test if $\widehat{\gamma}_{t-1}$ is equal to $\widehat{\varepsilon}_{t-1} I(\widehat{\varepsilon}_{t-1} < 0)$; and
- the positive sign bias test if $\widehat{\gamma}_{t-1}$ is equal to $\widehat{\varepsilon}_{t-1} I(\widehat{\varepsilon}_{t-1} > 0)$.

Instead of estimating three regressions to conduct these three tests, Engle and Ng (1993) advise to run a single regression accounting for all three effects:

$$\widehat{\varepsilon}_t^2 = \beta_0 + \beta_1 I\left(\widehat{\varepsilon}_{t-1} < 0\right) + \beta_2 \widehat{\varepsilon}_{t-1} I\left(\widehat{\varepsilon}_{t-1} < 0\right) + \beta_3 \widehat{\varepsilon}_{t-1} I\left(\widehat{\varepsilon}_{t-1} > 0\right) + \xi_t \qquad (3.12)$$

To implement the sign bias tests, we first have to obtain squared residuals from a model for the mean equation as follows:

```
tsset newdate
arima return, ar(1/2) nolog
predict residuals, residual
generate residuals2 = residuals^2
```

Now that we have obtained squared residuals, we are ready to run the regression comprising the three effects in (3.12).

```
. generate SBT = (residuals<0)

. generate NSBT = residuals*(residuals < 0)
(1 missing value generated)

. generate PSBT = residuals*(residuals > 0)
(1 missing value generated)

. regress residuals2 SBT NSBT PSBT
```

Source	SS	df	MS		
				Number of obs	= 16,102
				F(3, 16098)	= 5002.97
Model	.002058496	3	.000686165	Prob > F	= 0.0000
Residual	.002207867	16,098	1.3715e-07	R-squared	= 0.4825
				Adj R-squared	= 0.4824
Total	.004266363	16,101	2.6498e-07	Root MSE	= .00037

residuals2	Coef.	Std. Err.	t	P>|t|	[95% Conf. Interval]
SBT	-.0001555	7.90e-06	-19.68	0.000	-.0001709 -.00014
NSBT	-.0587697	.000542	-108.43	0.000	-.0598321 -.0577073
PSBT	.0345993	.0006078	56.93	0.000	.033408 .0357906
_cons	-.0001351	5.62e-06	-24.02	0.000	-.0001461 -.000124

The output shows that the coefficient associated with negative news (NSBT) is larger, in absolute terms, than the one associated with positive news (PSBT), suggesting that residuals effectively confirm the presence of a leverage effect.

We now present all the asymmetric GARCH models available.

3.5.1   SAARCH

The first asymmetric GARCH we introduce is the simple asymmetric ARCH (SAARCH) initially proposed by Engle (1990):

$$h_t = \omega + \alpha\varepsilon^2_{t-1} + \gamma\varepsilon_{t-1} + \beta h_{t-1}$$

where γ accounts for the leverage effect and is expected to take a negative sign, and the usual symmetric effect is captured by the coefficient α.

We can fit the SAARCH model by specifying the `saarch()` option in the `arch` command. Using this option, we are requiring the addition of an extra term to the standard ARCH and GARCH parameters, accounting for the leverage effect.

```
. arch return, arch(1) garch(1) saarch(1) nolog
initial values not feasible
(note:  default initial values infeasible; starting ARCH/ARMA estimates from 0)

ARCH family regression

Sample: 04jan1950 - 31dec2013                    Number of obs   =      16,102
Distribution: Gaussian                           Wald chi2(.)    =           .
Log likelihood =  54896.15                       Prob > chi2     =           .
```

return	Coef.	OPG Std. Err.	z	P>|z|	[95% Conf. Interval]	
return						
_cons	.0003048	.0000574	5.31	0.000	.0001923	.0004173
ARCH						
arch L1.	.0744891	.0017543	42.46	0.000	.0710507	.0779275
saarch L1.	-.000505	.0000168	-30.11	0.000	-.0005379	-.0004722
garch L1.	.9135125	.0022427	407.33	0.000	.9091169	.9179081
_cons	1.14e-06	7.07e-08	16.10	0.000	1.00e-06	1.28e-06

```
. estimates store SAARCH
```

As expected, the SAARCH coefficient enters the model with a negative sign, indicating that negative innovations $[\varepsilon_{t-1}I(\varepsilon_{t-1} < 0)]$ have a larger impact on volatility compared with positive ones.

3.5.2   TGARCH

The threshold GARCH (TGARCH), introduced by Zakoian (1994), is an asymmetric model specified in terms of absolute innovations rather than squared innovations (as in the standard GARCH(1,1)).

$$h_t = \omega + \alpha\left|\varepsilon_{t-1}\right| + \gamma\left|\varepsilon_{t-1}\right|I\left(\varepsilon_{t-1} > 0\right) + \beta h_{t-1} \tag{3.13}$$

where α accounts for the symmetric impact of innovations (the impact determined by the size of the news on the overall volatility, regardless of the sign of innovations) and γ is the coefficient accounting for the leverage effect, loading only positive news. We therefore expect γ to take a negative sign to recognize that positive news should have a smaller impact on variance than negative news.

We can fit a TGARCH model as follows:

```
. arch return, abarch(1) atarch(1) sdgarch(1) nolog

ARCH family regression

Sample: 04jan1950 - 31dec2013                    Number of obs   =      16,102
Distribution: Gaussian                           Wald chi2(.)    =           .
Log likelihood =  54927.15                       Prob > chi2     =           .
```

return	Coef.	OPG Std. Err.	z	P>|z|	[95% Conf. Interval]	
return						
_cons	.0002996	.0000563	5.33	0.000	.0001894	.0004099
ARCH						
abarch						
L1.	.1173896	.0018764	62.56	0.000	.1137119	.1210673
atarch						
L1.	-.0747301	.0021842	-34.21	0.000	-.079011	-.0704493
sdgarch						
L1.	.9230487	.0019728	467.89	0.000	.919182	.9269153
_cons	.00014	8.08e-06	17.33	0.000	.0001242	.0001559

```
. estimates store TGARCH
```

The `abarch(`*numlist*`)` option defines the absolute value of ARCH terms corresponding to α in (3.13), while the `atarch(`*numlist*`)` option identifies the absolute threshold ARCH accounting for the impact of positive news and corresponding to the γ term in (3.13). Finally, the `sdgarch(`*numlist*`)` option captures the heteroskedasticity effect, corresponding to β in (3.13).

Both parameters α and γ are statistically significant, with the leverage parameter γ taking a negative sign, exactly as expected.

3.5.3   GJR–GARCH

The GJR–GARCH model by Glosten, Jagannathan, and Runkle (1993) accounts for the leverage effect as follows:

$$h_t = \omega + \alpha\varepsilon_{t-1}^2 + \beta h_{t-1} + \gamma I\left(\varepsilon_{t-1} < 0\right)\varepsilon_{t-1}^2$$

where γ is the parameter accounting for the leverage effect that we expect to take a positive sign. When fitting the GJR model, we should pay attention when reading the output because the parameter γ is reported with a negative sign.

A limitation of the GJR model is that we must impose some bounds to ensure that volatility is positive, and these bounds make the estimation problem more difficult from a computational point of view.

We now fit the GJR model on the S&P 500 returns:

```
. arch return, arch(1) garch(1) tarch(1) nolog
initial values not feasible
(note:  default initial values infeasible; starting ARCH/ARMA estimates from 0)

ARCH family regression

Sample: 04jan1950 - 31dec2013                    Number of obs   =      16,102
Distribution: Gaussian                           Wald chi2(.)    =           .
Log likelihood =  54906.39                       Prob > chi2     =           .
```

return	Coef.	OPG Std. Err.	z	P>|z|	[95% Conf. Interval]	
return						
_cons	.0003258	.0000551	5.91	0.000	.0002178	.0004338
ARCH						
arch L1.	.1164146	.0024618	47.29	0.000	.1115896	.1212397
tarch L1.	-.0850406	.0031169	-27.28	0.000	-.0911496	-.0789316
garch L1.	.915121	.0022817	401.07	0.000	.9106489	.9195931
_cons	9.86e-07	6.18e-08	15.96	0.000	8.65e-07	1.11e-06

```
. estimates store GJR
```

In the output reported above, the leverage parameter γ takes a negative and statistically significant sign. We interpret this coefficient as evidence that bad news has a greater impact on returns than good news.

3.5.4 APARCH

Ding, Granger, and Engle (1993) propose the asymmetric power ARCH model (APARCH), corresponding to a Box–Cox function in the lagged innovations. The APARCH takes the following form:

$$h_t^{\frac{\delta}{2}} = \omega + \alpha \left(|\varepsilon_{t-1}| + \gamma\varepsilon_{t-1}\right)^{\delta} + \beta h_{t-1}^{\frac{\delta}{2}} \tag{3.14}$$

where $\delta > 0$ is the parameter playing the role of a Box–Cox transformation of h_t and γ captures the asymmetric effect. The parameter γ can take values between -1 and $+1$, although it is expected to be positive. The APARCH model in (3.14) nests several models:

- if $\gamma = 0$ and $\delta = 2$, APARCH reduces to GARCH;

- if $\delta = 2$, we obtain a model for the conditional variance; and

- if $\delta = 1$, we obtain a model for the conditional standard deviation.

For a detailed discussion on that point, see Ding, Granger, and Engle (1993).

Finally, the APARCH model works better if residuals are Student's t rather than normally distributed.

We can fit the APARCH model by specifying two options in the `arch` command, namely, `aparch(numlist)`, for the α and γ terms, and `pgarch(numlist)`, for the β term. We fit the APARCH model on the S&P 500 return time series.

```
. arch return, aparch(1) pgarch(1) nolog
ARCH family regression
Sample: 04jan1950 - 31dec2013                    Number of obs   =      16,102
Distribution: Gaussian                           Wald chi2(.)    =           .
Log likelihood =  54938.64                       Prob > chi2     =           .
```

return	Coef.	OPG Std. Err.	z	P>|z|	[95% Conf. Interval]	
return						
_cons	.0003036	.0000569	5.33	0.000	.000192	.0004151
ARCH						
aparch						
L1.	.0773795	.0026196	29.54	0.000	.0722451	.0825139
aparch_e						
L1.	-.4166682	.019523	-21.34	0.000	-.4549326	-.3784039
pgarch						
L1.	.9215671	.0021011	438.62	0.000	.9174491	.9256851
_cons	.0000287	6.84e-06	4.19	0.000	.0000153	.0000421
POWER						
power	1.317089	.045754	28.79	0.000	1.227413	1.406765

```
. estimates store APARCH
```

The term δ is equal to 1.32, suggesting that the dependent variable of our model is approximately the standard deviation. Parameter `aparch L1.` refers to coefficient α in (3.14), `aparch_e L1.` refers to γ, and `pgarch L1.` refers to β. The negative and statistically significant value assumed by coefficient γ confirms once again the presence of a leverage effect.

We propose an overall table to compare the models estimated in this section:

```
. estimates table SAARCH TGARCH GJR APARCH, stats(aic bic)
```

Variable	SAARCH	TGARCH	GJR	APARCH
return				
_cons	.00030478	.00029961	.00032579	.00030358
ARCH				
arch				
L1.	.07448914		.11641463	
saarch				
L1.	-.00050504			
garch				
L1.	.91351247		.91512099	
abarch				
L1.		.11738963		
atarch				
L1.		-.07473014		
sdgarch				
L1.		.92304866		
tarch				
L1.			-.0850406	
aparch				
L1.				.07737947
aparch_e				
L1.				-.41666823
pgarch				
L1.				.92156712
_cons	1.138e-06	.00014002	9.861e-07	.00002868
POWER				
power				1.3170891
Statistics				
aic	-109782.3	-109844.31	-109802.77	-109865.28
bic	-109743.87	-109805.88	-109764.34	-109819.16

This command is very useful because it allows us to compare many alternative models in a simple way. Here we request the parameters of the models listed in the command plus the AIC and the BIC. We can see that the constant of the mean equation is quite similar across all the models, being close to 0.0003. The same holds true for the GARCH parameter, being equal to 0.92 for the GJR and TGARCH parameters and to 0.91 for the SAARCH parameter. Finally, both the AIC and the BIC indicate that the APARCH model is the one providing the best trade-off between the goodness of fit and the number of parameters.

3.5.5   News impact curve

To evaluate the ability of a GARCH model to account for the leverage effect, we generally adopt the so-called news impact curve (NIC). The NIC measures how the volatility at time t reacts to news (error terms) released at time $t - 1$, assuming that information dated $t - 2$ and earlier remains constant, and with all lagged conditional variance evaluated at the level of the unconditional variance.

We can compute the news impact curve by using the `predict` command with the `at`($varname_\epsilon \mid \#_\epsilon\, varname_{\sigma^2} \mid \#_{\sigma^2}$) option, which allows us to obtain static predictions for a given set of disturbances.

The first step is to create a range of possible values for ε_t (for instance, between -4 and 4) by using the `generate` command:

```
. generate et = (_n-8000)/2000
```

We now fit one GARCH model and compute predictions for the conditional volatility h_t on the basis of parameter estimates and the vector of the possible innovations `et`. We consider a GARCH, a TGARCH, and a GJR model.

```
. * GARCH(1,1)
. arch return, arch(1) garch(1) nolog
ARCH family regression
Sample: 04jan1950 - 31dec2013                Number of obs   =      16,102
Distribution: Gaussian                       Wald chi2(.)    =           .
Log likelihood =  54784.48                   Prob > chi2     =           .
```

return	Coef.	OPG Std. Err.	z	P>|z|	[95% Conf. Interval]	
return						
_cons	.0004798	.0000537	8.93	0.000	.0003745	.000585
ARCH						
arch L1.	.0817023	.0016294	50.14	0.000	.0785086	.0848959
garch L1.	.9118854	.0021948	415.48	0.000	.9075837	.9161871
_cons	8.08e-07	6.70e-08	12.06	0.000	6.77e-07	9.39e-07

```
. predict varianceGARCH, variance at (et 1)
. label variable varianceGARCH "VarianceGARCH"
```

```
. * TGARCH(1,1)
. arch return, abarch(1) atarch(1) sdgarch(1) nolog
ARCH family regression
Sample: 04jan1950 - 31dec2013                Number of obs   =      16,102
Distribution: Gaussian                       Wald chi2(.)    =           .
Log likelihood =  54927.15                   Prob > chi2     =           .
```

return	Coef.	OPG Std. Err.	z	P>|z|	[95% Conf. Interval]	
return						
_cons	.0002996	.0000563	5.33	0.000	.0001894	.0004099
ARCH						
abarch						
L1.	.1173896	.0018764	62.56	0.000	.1137119	.1210673
atarch						
L1.	-.0747301	.0021842	-34.21	0.000	-.079011	-.0704493
sdgarch						
L1.	.9230487	.0019728	467.89	0.000	.919182	.9269153
_cons	.00014	8.08e-06	17.33	0.000	.0001242	.0001559

```
. predict varianceTGARCH, variance at (et 1)

. label variable varianceTGARCH "VarianceTGARCH"

. * GJR(1,1)
. arch return, arch(1) garch(1) tarch(1) nolog
initial values not feasible
(note:  default initial values infeasible; starting ARCH/ARMA estimates from 0)
ARCH family regression
Sample: 04jan1950 - 31dec2013                Number of obs   =      16,102
Distribution: Gaussian                       Wald chi2(.)    =           .
Log likelihood =  54906.39                   Prob > chi2     =           .
```

return	Coef.	OPG Std. Err.	z	P>|z|	[95% Conf. Interval]	
return						
_cons	.0003258	.0000551	5.91	0.000	.0002178	.0004338
ARCH						
arch						
L1.	.1164146	.0024618	47.29	0.000	.1115896	.1212397
tarch						
L1.	-.0850406	.0031169	-27.28	0.000	-.0911496	-.0789316
garch						
L1.	.915121	.0022817	401.07	0.000	.9106489	.9195931
_cons	9.86e-07	6.18e-08	15.96	0.000	8.65e-07	1.11e-06

```
. predict varianceGJR, variance at (et 1)
```

```
. label variable varianceGJR "VarianceGJR"
. line varianceGARCH varianceTGARCH varianceGJR et in 2/1, m(i) c(1)
> title(News response function)
```

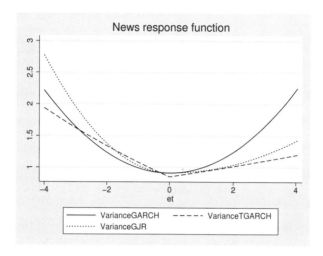

Figure 3.8. News impact curve for GARCH, TGARCH, and GJR

The shape of the news impact curve in figure 3.8 highlights the ability of GJR and TGARCH to account for the greater impact of negative news versus positive news. In fact, we can see that the simple GARCH model is not able to treat positive and negative news differently.

3.5.6 Forecasting comparison

We now want to compare the alternative models in terms of their capability to forecast volatility out of sample. Because a true measure of volatility does not exist, we need to use its latent value. Patton (2011) identifies a class of loss functions that asymptotically generate the same ranking of models, regardless of the volatility estimator adopted. Patton (2011) shows that two loss functions belong to this class, the mean squared error (MSE) and the quasilikelihood (QLIKE), respectively defined as

$$\text{MSE} : L\left(\sigma^2, h\right) = \left(\sigma^2 - h\right)^2$$
$$\text{QLIKE} : L\left(\sigma^2, h\right) = \log h + \frac{\sigma^2}{h}$$

where σ^2 is the true latent variance and h is a variance forecast.

The QLIKE loss is generally less affected by outliers, while MSE is sensitive both to extreme observations in the sample and to the volatility of returns. Therefore, in our exercise, we are going to adopt the QLIKE loss function. We conduct our comparison on the three GARCH models we analyzed in the previous subsection: GARCH, TGARCH, and

GJR. We must first obtain forecasts for the volatility by using the command `predict` with the `variance` option and the `dynamic()` option to indicate where the forecast should start. We fit the alternative models on the period 1 January 2013 to 30 November 2013, leaving out one month for the out-of-sample comparison; see figure 3.9.

```
. * GARCH
. arch return if date >= mdy(01,01,2013) & date < = mdy(11,30,2013), arch(1)
> garch(1)
  (output omitted)
. predict var_GARCH_Forecast if date  >= mdy(01,01,2013), variance
> dynamic(mdy(11,30,2013))
(15,851 missing values generated)
. * TGARCH
. arch return if date >= mdy(01,01,2013) & date < = mdy(11,30,2013), abarch(1)
> atarch(1) sdgarch(1) nolog
  (output omitted)
. predict var_TGARCH_Forecast if date  >= mdy(01,01,2013), variance
> dynamic(mdy(11,30,2013))
(15,851 missing values generated)
. * GJR
. arch return if date >= mdy(01,01,2013) & date < = mdy(11,30,2013), arch(1)
> garch(1) tarch(1) nolog
  (output omitted)
. predict var_GJR_Forecast if date  >= mdy(01,01,2013), variance
> dynamic(mdy(11,30,2013))
(15,851 missing values generated)
. generate return2 = return^2
(1 missing value generated)
. tsset date
        time variable: date, 1/3/1950 to 12/31/2013, but with gaps
                delta: 1 day
. tsline return2 var_GARCH_Forecast var_TGARCH_Forecast
> var_GJR_Forecast if date>=mdy(01,01,2013), tline(30nov2013)
```

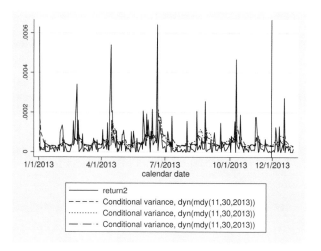

Figure 3.9. Variance forecast comparison

```
. * QL criteria
. generate QLvar_GARCH =
> sum((return2/var_GARCH_Forecast-ln(return2/var_GARCH_Forecast)-1))
> if date>=mdy(12,01,2013)
(16,082 missing values generated)

. generate QLvar_TGARCH =
> sum((return2/var_TGARCH_Forecast-ln(return2/var_TGARCH_Forecast)-1))
> if date>=mdy(12,01,2013)
(16,082 missing values generated)

. generate QLvar_GJR =
> sum((return2/var_GJR_Forecast-ln(return2/var_GJR_Forecast)-1))
> if date>=mdy(12,01,2013)
(16,082 missing values generated)

. summarize QLvar_GARCH QLvar_TGARCH QLvar_GJR
```

Variable	Obs	Mean	Std. Dev.	Min	Max
QLvar_GARCH	21	16.82193	12.16008	.6447565	37.84296
QLvar_TGARCH	21	15.63339	10.60684	.8129278	34.17938
QLvar_GJR	21	15.58642	10.9251	.6947053	34.56509

```
. tsset newdate
        time variable:  newdate, 03jan1950 to 31dec2013
                delta:  1 day
```

The quasilikelihood variance statistic is equal to 37.84 for the GARCH model, 34.18 for the TGARCH, and 34.57 for the GJR, suggesting that the TGARCH model provides the best forecasting performance of volatility because it minimizes the loss function quasilikelihood variance.

3.6   Alternative GARCH models

In this section, we present some alternative GARCH models that do not account for the leverage effect. The power ARCH model is a class of models where the dependent variable is a transformation of the variance. The nonlinear ARCH models allow for the presence of some constants around which errors can move.

3.6.1   PARCH

We can express the power ARCH (PARCH) model by Higgins and Bera (1992) as

$$h_t = \omega + \alpha \varepsilon_{t-1}^{\varphi} + \beta h_{t-1}^{\varphi/2} \tag{3.15}$$

where $\alpha \varepsilon_{t-1}^{\varphi}$ is the PARCH term and $\beta h_{t-1}^{\varphi/2}$ identifies the PGARCH term.

To fit a PARCH model, we use the **arch** command plus the **parch**(*numlist*) option for the PARCH term and the **pgarch**(*numlist*) option for the PGARCH term. We now fit a PARCH(1,1) model for our S&P 500 return time series:

```
. arch return, parch(1) pgarch(1) nolog

ARCH family regression

Sample: 04jan1950 - 31dec2013                    Number of obs   =      16,102
Distribution: Gaussian                           Wald chi2(.)    =           .
Log likelihood =  54795.87                       Prob > chi2     =           .
```

return	Coef.	OPG Std. Err.	z	P>|z|	[95% Conf. Interval]
return						
_cons	.000488	.0000531	9.19	0.000	.0003839	.0005921
ARCH						
parch L1.	.090654	.0025948	34.94	0.000	.0855683	.0957396
pgarch L1.	.9143575	.0022262	410.73	0.000	.9099943	.9187207
_cons	7.53e-06	2.39e-06	3.15	0.002	2.84e-06	.0000122
POWER						
power	1.544414	.0608173	25.39	0.000	1.425214	1.663614

In the output above, the **power** parameter corresponds to the coefficient φ in (3.15), which in our case takes the value 1.54. Thus, the model we are fitting lies between a model on squared innovations and variance (the case when φ is 2) and a model on simple innovations and volatility (the case when φ is 1).

3.6.2 NGARCH

The models presented so far rely on the assumption that "no news is good news", or in other words, only the arrival of news can have an impact on the market. Sometimes even the lack of new information can contribute in raising some nervousness in the markets, making prices move in a strong way and leading to a raise in volatility.

Nonlinear GARCH (NGARCH) models explicitly deal with this feature. From a mathematical point of view, NGARCH models recognize this feature by allowing the impact of innovations to be centered around some constants instead of around 0.

The nonlinear ARCH model enters the model for conditional variance with nonlinear ARCH terms, by adding multiple shifts $k_1, k_2, \ldots, k_n$. For instance, we can write an NGARCH(2) model as

$$h_t = \omega + \alpha_1 \left(\varepsilon_{t-1} - k_1\right)^2 + \alpha_2 \left(\varepsilon_{t-2} - k_2\right)^2 + \beta h_{t-1} \tag{3.16}$$

From (3.16), it is evident that even when there is no news—that is, when ε_t equals 0—the α_1 and α_2 are still associated with some values k_1 and k_2, respectively. The model implicitly recognizes that the lack of news could raise the uncertainty in the markets, determining an increase in volatility.

We can fit the NGARCH model by specifying the **narch**(*numlist*) option in the usual **arch** command. We also add the **arch0(xbwt)** option to modify the priming values used by default. We are now using the weighted sum of the squared residuals from the current conditional mean equation, which places more weight at the beginning of the sample, instead of the default expected unconditional variance.

```
. arch return, narch(1/2) garch(1) arch0(xbwt) nolog
ARCH family regression
Sample: 04jan1950 - 31dec2013                    Number of obs   =      16,102
Distribution: Gaussian                           Wald chi2(.)    =           .
Log likelihood =  54917.95                       Prob > chi2     =           .
```

return	Coef.	OPG Std. Err.	z	P>|z|	[95% Conf. Interval]	
return						
_cons	.0003549	.0000574	6.18	0.000	.0002424	.0004674
ARCH						
narch						
L1.	.0988348	.0035914	27.52	0.000	.0917957	.1058738
L2.	-.0368741	.0045908	-8.03	0.000	-.045872	-.0278762
narch_k						
L1.	.0051191	.0003178	16.11	0.000	.0044963	.005742
L2.	.0083643	.0012223	6.84	0.000	.0059686	.01076
garch						
L1.	.9295937	.0025149	369.63	0.000	.9246645	.9345228
_cons	7.87e-07	3.22e-07	2.44	0.015	1.55e-07	1.42e-06

From the output reported above, we can note that both coefficients k_1 and k_2, identified by the label narch_k, are positive and statistically different from 0, suggesting that the statement "no news is good news" does not always hold true. With the label narch, Stata denotes the coefficients α_1 and α_2 in (3.16).

3.6.3 NGARCHK

We can modify the model in (3.16) forcing all the shifts k_i to take the same value k_0. This new model, the NGARCHK, is a nonlinear ARCH with one shift. The specification with two lags now takes the form

$$h_t = \omega + \beta h_{t-1} + \alpha_1 \left(\varepsilon_{t-1} - k_0\right)^2 + \alpha_2 \left(\varepsilon_{t-1} - k_0\right)^2$$

We can fit this model by specifying the `narchk()` option with the `arch` command:

```
. arch return, narchk(1/2) garch(1) nolog

ARCH family regression
Sample: 04jan1950 - 31dec2013                    Number of obs   =      16,102
Distribution: Gaussian                           Wald chi2(.)    =           .
Log likelihood =  54912.25                       Prob > chi2     =           .
```

	Coef.	OPG Std. Err.	z	P>|z|	[95% Conf.	Interval]
return						
_cons	.000327	.0000572	5.72	0.000	.000215	.0004391
ARCH						
narch						
L1.	.110976	.0036909	30.07	0.000	.1037418	.1182101
L2.	-.0497279	.004272	-11.64	0.000	-.0581008	-.041355
narch_k	.0035171	.0001543	22.80	0.000	.0032147	.0038194
garch						
L1.	.9299127	.002365	393.20	0.000	.9252774	.9345479
_cons	8.09e-08	6.19e-08	1.31	0.191	-4.05e-08	2.02e-07

In the output reported above, we see that both **narch** coefficients are statistically significant, with the first lag taking a positive sign and the second lag taking a negative sign. These two coefficients are quite similar to those we obtained when we fit the NGARCH model, but here we just have one k coefficient (**narch_k** equal to 0.93), while the NGARCH model required two k coefficients.

4 Multivariate GARCH models

4.1 Introduction

Modeling returns volatility has received a great deal of attention since the introduction of GARCH models. Although it is crucial to properly describe and capture the dynamic of volatility, in empirical applications we usually deal with a portfolio of assets, making it important to model the joint dynamics of the assets composing the portfolio and therefore involving both volatilities and correlations. For instance, when taking decisions on portfolio optimization, we need to consider the dynamic of all the assets included in the portfolio to properly deal with risk diversification. Similarly, when assessing the risk of a portfolio, we need to account for the correlation of assets.

Thus, it is straightforward to extend the univariate GARCH models described in chapter 3 to the multivariate GARCH (MGARCH) models, which have seen interesting developments over the past decade. The use of MGARCH models is quite wide in finance: Bollerslev, Engle, and Wooldridge (1988) and Hansson and Hordahl (1998) illustrate some portfolio applications, while Tse and Tsui (2002) and Bae, Karolyi, and Stulz (2003) adopt MGARCH models to investigate volatility and correlation transmission as well as spillover effects in studies of contagion. A good reference for multivariate volatility models is Silvennoinen and Teräsvirta (2009).

To formally introduce MGARCH, it is useful to provide a short introduction to vector autoregressive (VAR) process models. The VAR model is the natural multivariate extension of the univariate AR process. In the VAR model, all the variables of interest are endogenous, and it is therefore necessary to model all the variables simultaneously.

A VAR model with p terms and three variables x, y, and z can be written as follows:

$$
\begin{aligned}
y_t &= \rho_{11}y_{t-1} + \rho_{12}y_{t-2} + \cdots + \rho_{1p}y_{t-p} + \beta_{11}x_{t-1} + \beta_{12}x_{t-2} + \cdots + \beta_{1p}x_{t-p} + \\
&\quad \gamma_{11}z_{t-1} + \gamma_{12}z_{t-2} + \cdots + \gamma_{1p}z_{t-p} + \varepsilon_{1t} \\
x_t &= \rho_{21}y_{t-1} + \rho_{22}y_{t-2} + \cdots + \rho_{2p}y_{t-p} + \beta_{21}x_{t-1} + \beta_{22}x_{t-2} + \cdots + \beta_{2p}x_{t-p} + \\
&\quad \gamma_{21}z_{t-1} + \gamma_{22}z_{t-2} + \cdots + \gamma_{2p}z_{t-p} + \varepsilon_{2t} \\
z_t &= \rho_{31}y_{t-1} + \rho_{32}y_{t-2} + \cdots + \rho_{3p}y_{t-p} + \beta_{31}x_{t-1} + \beta_{32}x_{t-2} + \cdots + \beta_{3p}x_{t-p} + \\
&\quad \gamma_{31}z_{t-1} + \gamma_{32}z_{t-2} + \cdots + \gamma_{3p}z_{t-p} + \varepsilon_{3t}
\end{aligned}
$$

In matrix notation, this is written as

$$
Y_t = V + \Phi_1 Y_{t-1} + \Phi_2 Y_{t-2} + \cdots + \Phi_p Y_{t-p} + \Xi_t \tag{4.1}
$$

where $Y_t = (y_{1t}, \ldots, y_{mt})$ vector of m variables at time t; $\Phi_1, \ldots, \Phi_p$ is a matrix of $m \times m$ coefficients; $V = (v_1, \ldots, v_m)$ vector of m constants; and $\Xi_t = (\varepsilon_{1t}, \ldots, \varepsilon_{mt})$ vector of m random errors following a multivariate white-noise process $(0, \Sigma)$.

The process in (4.1) is covariance stationary if

$$\det (I_k - \Phi_1 z - \cdots - \Phi_p z^p) \neq 0 \quad \text{for } |z| < 1$$

We can always express a VAR with p terms as a VAR with one term by setting

$$\widetilde{y}_t = (y_t \; y_{t-1} \; \ldots \; y_{t-p+1})$$

$$\widetilde{v} = (v \, 0 \, \ldots \, 0)$$

$$\widetilde{\Phi} = \begin{bmatrix} \phi_1 & \phi_2 & \ldots & \phi_{p-1} & \phi_p \\ I_k & 0 & \ldots & 0 & 0 \\ 0 & I_k & \ldots & 0 & 0 \\ \ldots & \ldots & \ldots & \ldots & \ldots \\ 0 & 0 & \ldots & I_k & 0 \end{bmatrix}$$

$$\widetilde{\Xi}_t = (\varepsilon_t \, 0 \, \ldots \, 0)$$

Then it is possible to rewrite a VAR(p) process as a VAR(1) process:

$$\widetilde{y}_t = \widetilde{v} + \widetilde{\Phi}\widetilde{y}_{t-1} + \widetilde{\Xi}_t$$

4.2   Multivariate GARCH

Recalling (4.1), we can model innovations as

$$\Xi_t = H_t^{1/2}(\theta)Z_t \tag{4.2}$$

where $Z_t(m \times 1)$ is a Gaussian white-noise process with a mean vector of 0 and an identity variance–covariance matrix I_m. $H_t^{1/2}(m \times m)$ is a lower triangular matrix with a positive principal diagonal that is obtained by the Cholesky factorization of matrix H_t, which is the conditional variance–covariance matrix of VAR residuals.

Let us compute the expected value for the innovations in (4.2),

$$E\left(\Xi_t\right) = E\left(\sqrt{H_t}Z_t\right)$$

$$E\left(\Xi_t\right) = \sqrt{H_t}E\left(Z_t\right)$$

$$E\left(\Xi_t\right) = 0$$

and the variance,

$$E\left(\Xi_t\Xi_t'\right) = E\left(\sqrt{H_t}Z_tZ_t'\sqrt{H_t'}\right)$$

$$E\left(\Xi_t\Xi_t'\right) = \sqrt{H_t}E\left(Z_tZ_t'\right)\sqrt{H_t'}$$

$$E\left(\Xi_t\Xi_t'\right) = \sqrt{H_t}I_m\sqrt{H_t'}$$

$$E\left(\Xi_t\Xi_t'\right) = I_mH_t$$

$$E\left(\Xi_t\Xi_t'\right) = H_t$$

We have shown that residuals follow a multivariate normal distribution with an expected value equal to a null vector and a variance–covariance equal to H_t. The main aim of MGARCH models is to specify a model for H_t that best fits the real dynamic of variances and covariances.

We face some important issues when dealing with MGARCH models. In general, there is a trade-off between the parsimony and the flexibility that alternative MGARCH specifications can offer. The first generation of MGARCH models is characterized by a high degree of flexibility with difficulties in model estimation, even when few series are taken into consideration. On the other hand, MGARCH specifications require a small number of parameters, thus making it quick to be estimated but at the cost of losing some flexibility.

Another issue with MGARCH models is the way in which the positive semidefiniteness of the conditional variance–covariance matrix is guaranteed. We discuss this issue at length when presenting the alternative specifications.

Several alternative MGARCH specifications exist. In particular, Bauwens, Laurent, and Rombouts (2006) distinguish three nonmutually exclusive approaches for constructing multivariate GARCH models:

1. direct generalizations of the univariate GARCH model of Bollerslev (1986);

2. linear combinations of univariate GARCH models; and

3. nonlinear combinations of univariate GARCH models.

See also Silvennoinen and Teräsvirta (2009) for a survey on MGARCH models. In this chapter, we focus on the first and the third approaches listed above.

4.3   Direct generalizations of the univariate GARCH model of Bollerslev

The models belonging to this category can be seen as multivariate extensions of the univariate GARCH model (Bollerslev 2010). The main issue with these models is that they suffer from the course of dimensionality, similarly to VAR models when modeling the conditional mean of a number of time series. The number of parameters we have to estimate rapidly increases as the number of time series considered gets larger, making these models unsuitable for empirical applications dealing with even a small number of time series. Furthermore, these models require added constraints to ensure that the variance–covariance matrix H_t is positive semidefinite. Aside from these two limits, models belonging to this category provide great flexibility, and so we can still use them for empirical applications involving few time series.

To this category belong the following models:

1. vech models,

2. BEKK models, and

3. factor models (F-GARCH).

Because multivariate GARCH models relying on factor decomposition (F-GARCH) are not implemented in Stata, we will not illustrate them in this book.

4.3.1   Vech model

To describe vech models, we start by introducing the vech operator. The vech operator transforms a symmetric matrix $A(m \times m)$ in a vector of dimension $m(m+1)/2$, lining up the m column vectors that form the matrix A, excluding elements above the principal diagonal. See the following simple example:

$$
\text{vech} \begin{bmatrix} a_{11} & a_{12} & a_{13} \\ a_{21} & a_{22} & a_{23} \\ a_{31} & a_{32} & a_{33} \end{bmatrix} = \begin{bmatrix} a_{11} \\ a_{21} \\ a_{31} \\ a_{22} \\ a_{32} \\ a_{33} \end{bmatrix}
$$

The vech specification is proposed by Bollerslev, Engle, and Wooldridge (1988). In this formulation, every element of the conditional covariance matrix is a function of the lagged conditional covariance matrix, a function of lagged squared standardized residuals, and cross-products of standardized residuals.

A vech model with (p, q) terms can be specified as follows:

$$
\text{vech}(H_t) = \Omega + \sum_{i=1}^{p} A_i \text{vech}\left(\Xi_{t-i}\Xi'_{t-i}\right) + \sum_{j=1}^{q} B_j \text{vech}\left(H_{t-j}\right) \tag{4.3}
$$

where Ω is a constant vector with $m(m+1)/2$ elements, and A_i and B_j are matrices of dimension $[m(m+1)/2, m(m+1)/2]$.

To ensure that the process is covariance stationary, we need to impose the condition that the eigenvalues of the matrix $A + B$ lie between -1 and 1. This requirement implies that at every step, we need to compute the eigenvalues of the matrices A and B, a process that makes the estimation of the model quite slow.

To clarify this point, consider the case with $p = q = 1$ and for two series, $m = 2$:

$$\text{vech} \begin{bmatrix} h_{11,t} & h_{12,t} \\ h_{21,t} & h_{22,t} \end{bmatrix} = \begin{bmatrix} \omega_1 \\ \omega_2 \\ \omega_3 \end{bmatrix} + \begin{bmatrix} a_{11} & a_{12} & a_{13} \\ a_{21} & a_{22} & a_{23} \\ a_{31} & a_{32} & a_{33} \end{bmatrix}$$

$$\text{vech} \begin{bmatrix} \varepsilon_{1,t}^2 & \varepsilon_{1,t-1}\varepsilon_{2,t-1} \\ \varepsilon_{2,t-1}\varepsilon_{1,t-1} & \varepsilon_{2,t}^2 \end{bmatrix} + \begin{bmatrix} b_{11} & b_{12} & b_{13} \\ b_{21} & b_{22} & b_{23} \\ b_{31} & b_{32} & b_{33} \end{bmatrix}$$

$$\text{vech} \begin{bmatrix} h_{11,t-1} & h_{12,t-1} \\ h_{21,t-1} & h_{22,t-1} \end{bmatrix}$$

$$\begin{bmatrix} h_{11,t} \\ h_{21,t} \\ h_{22,t} \end{bmatrix} = \begin{bmatrix} \omega_1 \\ \omega_2 \\ \omega_3 \end{bmatrix} + \begin{bmatrix} a_{11} & a_{12} & a_{13} \\ a_{21} & a_{22} & a_{23} \\ a_{31} & a_{32} & a_{33} \end{bmatrix} \begin{bmatrix} \varepsilon_{1,t}^2 \\ \varepsilon_{2,t-1}\varepsilon_{1,t-1} \\ \varepsilon_{2,t}^2 \end{bmatrix} + $$

$$\begin{bmatrix} b_{11} & b_{12} & b_{13} \\ b_{21} & b_{22} & b_{23} \\ b_{31} & b_{32} & b_{33} \end{bmatrix} \begin{bmatrix} h_{11,t-1} \\ h_{21,t-1} \\ h_{22,t-1} \end{bmatrix}$$

This simple example is useful to get a grasp about the issue of dimensionality affecting vech models. By considering just two series and just one lag, we need to estimate 21 parameters. The number of parameters to be estimated for a general case with m time series and lags p and q is $m(m+1)/2 + (p+q)\{m(m+1)/2\}^2$. For instance, by setting $p = q = 1$ and $m = 5$, the overall number of parameters is 465, which is an unreasonable number for modeling just five time series!

Under the assumption that standardized residuals follow a multivariate normal distribution, we can estimate the vech model by maximum likelihood, where the log-likelihood function takes the following form:

$$\text{LogL}(\theta) = -0.5Tm \log(2\pi) - 0.5 \sum_{t=1}^{T} \left[\log \{ \det(H_t) \} + \zeta_t H_t^{-1} \zeta_t' \right] \tag{4.4}$$

where ζ_t is a row vector of residuals from a VAR model fit on returns. If we do not specify any model for returns, then we have simply $\zeta_t = r_t$.

When we assume the multivariate Student's t distribution with v degrees of freedom, we come up with the following log-likelihood function:

$$\text{LogL}(\theta) = T \log \left\{ \Gamma \left(\frac{v+m}{2} \right) \right\} - T \log \left\{ \Gamma \left(\frac{v}{2} \right) \right\} - T \frac{m}{2} \log \left\{ (v-2)\pi \right\} -$$

$$0.5 \sum_{t=1}^{T} \left[\log \left\{ \det (H_t) \right\} + \frac{v+m}{2} \log \left(1 + \frac{\zeta_t H_t^{-1} \zeta_t'}{v-2} \right) \right] \quad (4.5)$$

From the two log-likelihood functions (4.4) and (4.5), we can see that the maximization procedure requires the inversion of the conditional covariance matrix H_t at each time/iteration t. The inversion of the matrix H_t contributes to dramatically slow down the maximization procedure.

The vech model is flexible because it allows each variance and covariance to follow its own dynamics, which in turn is a function of p lags of standardized residuals and of q lags of the variances and covariances. The vech model can achieve its high flexibility at the cost of a huge number of parameters. Moreover, to ensure the positive semidefiniteness of H_t, we need to impose several constraints that are neither simple nor easily implementable. A final issue of this model is that parameters of matrices Ω, A, and B in (4.3) are not uniquely identified. To make the model feasible when working with a reasonable number of time series, it is necessary to impose some constraints in such a way to reduce the number of parameters we need to estimate and simplify the conditions we have to impose for ensuring that H_t is positive semidefinite.

The vech model and the other MGARCH models are all different from each other in how they deal with flexibility, allowing us to describe in a richer way the time dynamics of the variance–covariance matrix and ensuring that the estimation process does not become cumbersome.

4.3.2   Diagonal vech model

Bollerslev, Engle, and Wooldridge (1988) introduced the diagonal vech (dvech) model. The dvech relies on the assumption that matrices A_i and B_j in (4.3) are diagonal. Imposing those restrictions implies that every variance $h_{ii,t}$ depends only on $\varepsilon_{i,t-z}^2$, for $z = 1, \ldots, p$, and on its lag values and that each covariance $h_{ij,t}$ (with $i \neq j$) depends only on $(\varepsilon_{i,t-z}, \varepsilon_{j,t-z})$, for $z = 1, \ldots, p$, and on its lag values.

Under these conditions, Bollerslev, Engle, and Nelson (1994) show that it is possible to obtain conditions ensuring that the matrix H_t is positive semidefinite for all t. Let us consider the case of two time series, $m = 2$, and just one lag, $p = q = 1$. The dvech model takes the following form:

$$
\begin{bmatrix} h_{11,t} \\ h_{21,t} \\ h_{22,t} \end{bmatrix} = \begin{bmatrix} \omega_1 \\ \omega_2 \\ \omega_3 \end{bmatrix} + \begin{bmatrix} a_{11} & 0 & 0 \\ 0 & a_{22} & 0 \\ 0 & 0 & a_{33} \end{bmatrix} \begin{bmatrix} \varepsilon_{1,t}^2 \\ \varepsilon_{2,t-1}\varepsilon_{1,t-1} \\ \varepsilon_{2,t}^2 \end{bmatrix} +
$$
$$
\begin{bmatrix} b_{11} & 0 & 0 \\ 0 & b_{22} & 0 \\ 0 & 0 & b_{33} \end{bmatrix} \begin{bmatrix} h_{11,t-1} \\ h_{21,t-1} \\ h_{22,t-1} \end{bmatrix}
$$

This dvech model is simpler than the vech model considered before because here we just have to estimate 9 parameters versus the 21 of the previous model. In the dvech example just reported, we considered only two time series, $m = 2$. When instead we consider a number m of time series with lags p and q, the overall number of parameters we have to estimate is $m(m + 1)/2(1 + p + q)$. Therefore, setting $p = q = 1$ and $m = 5$, we have to estimate 45 parameters, which is far below the 465 required by the vech model.

We still need to impose some constraints to ensure that H_t is positive semidefinite, although we can set them in an easier way than the traditional vech model requires. Finally, the assumption that each variance and covariance depends only on its lags and on its error lags is quite restrictive.

4.3.3   BEKK model

An MGARCH representation that has a rich structure of dependence and at the same time guarantees the positive semidefiniteness of H_t without imposing any conditions is the BEKK model, so named because its authors are Baba, Engle, Kraft, and Kroner (1991). The BEKK model can be seen as a restricted version of the vech model with the great advantage of requiring a lower number of estimates. We can represent the BEKK(p, q) model as follows:

$$
H_t = C'C + \sum_{i=1}^{p} A_i' \Xi_{t-i} \Xi_{t-i}' A_i + \sum_{j=1}^{q} B_j' H_{t-j} B_j \tag{4.6}
$$

where $C(m \times m)$ is an upper triangular matrix with a positive principal diagonal and full rank, and A_i and B_j $(m \times m)$ are coefficient matrices.

A great advantage of the BEKK model over the vech model is that the matrix $C'C$ is positive semidefinite by construction, ensuring in turn that even the matrices $H_0, H_{t-1}, \ldots, H_{t-q}$ are positive semidefinite. Thus, the maximization problem is an unconstrained one, requiring a lower computational load compared with the constrained problems characterizing vech models.

In the literature, two simplified versions of the BEKK model were proposed. The first one is the diagonal BEKK model, where both A and B matrices in (4.6) are diagonal, implying that each variance and covariance depends just on its own lags and on its own lagged innovations. The second model is the scalar BEKK model, where A and B are scalar matrices—that is, $A = aI$ and $B = bI$—implying that each variance and covariance follows the same pattern governed by the same coefficient.

4.3.4 Empirical application

Data description

In our empirical example, we consider daily log-returns for the years 1991–2013 for four stock indexes, namely, S&P 500, FTSE 100, Nikkei 225, and DAX 30. We start our analysis by describing the time series of log-returns for the four time series considered. We proceed as follows:

```
. use http://www.stata-press.com/data/feus/index

. tsset date
        time variable:  date, 1/2/1991 to 12/30/2013, but with gaps
                delta:  1 day

. tsline Dax30, name(Dax30)

. tsline FTSE100, name(FTSE100)

. tsline Nikkei225, name(Nikkei225)

. tsline SP500, name(SP500)

. graph combine Dax30 FTSE100 Nikkei225 SP500
```

We obtain each individual chart, and then we put them all together by using the `graph combine` command. We report the final output in figure 4.1.

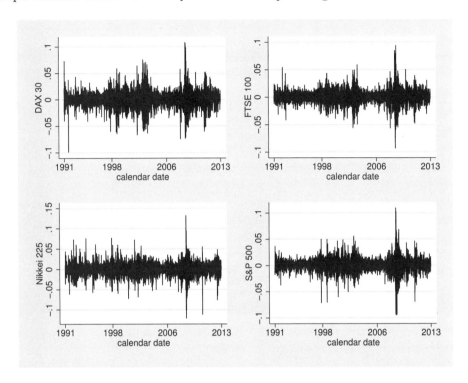

Figure 4.1. Daily returns for S&P 500, FTSE 100, Nikkei 225, and DAX 30, for the years 1991–2013

From figure 4.1, we can easily identify periods of high volatility followed by periods of lower volatility. In particular, we can see that all the indexes present a substantial increase of volatility in correspondence to the burst of the financial crisis in 2008.

To get some useful insights about the dataset considered, we run the `summarize` command on the four series.

```
. summarize Dax30 FTSE100 Nikkei225 SP500, detail
                              DAX 30
```

	Percentiles	Smallest		
1%	-.0432093	-.0987092		
5%	-.0229306	-.0743346		
10%	-.0158922	-.0733552	Obs	5,828
25%	-.0064202	-.0727027	Sum of Wgt.	5,828
50%	.0007867		Mean	.0003297
		Largest	Std. Dev.	.014443
75%	.0076764	.0755268		
90%	.015581	.0984283	Variance	.0002086
95%	.0219517	.1068509	Skewness	-.1046713
99%	.0379893	.1079747	Kurtosis	7.763967

```
                             FTSE 100
```

	Percentiles	Smallest		
1%	-.0315471	-.0926456		
5%	-.0172844	-.0817844		
10%	-.0120624	-.0742865	Obs	5,828
25%	-.0053322	-.0587007	Sum of Wgt.	5,828
50%	.0001617		Mean	.0001891
		Largest	Std. Dev.	.0113336
75%	.0060211	.0774297		
90%	.0121098	.0793676	Variance	.0001285
95%	.0171552	.0846914	Skewness	-.1085693
99%	.0303217	.0938424	Kurtosis	9.22138

```
                            Nikkei 225
```

	Percentiles	Smallest		
1%	-.0400613	-.1211103		
5%	-.0234349	-.1115343		
10%	-.0169184	-.1011604	Obs	5,828
25%	-.0076426	-.1008796	Sum of Wgt.	5,828
50%	0		Mean	-.0000792
		Largest	Std. Dev.	.014918
75%	.0077592	.0754969		
90%	.0168958	.0765533	Variance	.0002225
95%	.0224853	.0949415	Skewness	-.2238432
99%	.0370382	.1323458	Kurtosis	8.588183

```
                              S&P 500
```

	Percentiles	Smallest		
1%	-.0317961	-.0946951		
5%	-.0176624	-.0935366		
10%	-.0118314	-.0921896	Obs	5,828
25%	-.0045195	-.079224	Sum of Wgt.	5,828
50%	.0003358		Mean	.0002885
		Largest	Std. Dev.	.0115444
75%	.0056386	.0669226		
90%	.0119207	.0683664	Variance	.0001333
95%	.0167572	.1024573	Skewness	-.2401342
99%	.0330068	.109572	Kurtosis	12.04083

The four indexes show similar means and standard deviations. They differ more with respect to the kurtosis, with the S&P 500 showing a higher value than the other three indexes.

We now evaluate the autocorrelogram of the four series to identify the proper specification for the mean equation. We report the autocorrelogram in figure 4.2.

```
ac Dax30, name(ACDax30)
ac FTSE100, name(ACFTSE100)
ac Nikkei225, name(ACNikkei225)
ac SP500, name(ACSP500)
graph combine ACDax30 ACFTSE100 ACNikkei225 ACSP500
```

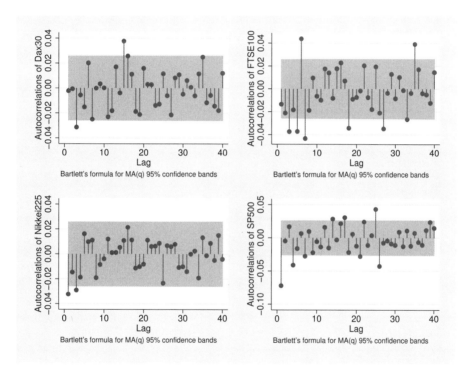

Figure 4.2. Autocorrelogram for daily returns of S&P 500, FTSE 100, Nikkei 225, and DAX 30

We do not find a clear pattern in the autocorrelogram suggesting the number of terms to be included for an ARMA model. Therefore, we just adopt an autoregressive model with one term for each of them.

Dvech model

We start our illustration of the MGARCH models by fitting the dvech model on the four indexes considered. However, because this model can be cumbersome to fit, we start from a restricted version of it. With this purpose, we start by specifying some constraints requiring all the ARCH coefficients to be equal among them, and afterward we do the same for the GARCH part.

```
. tsset newdate
        time variable:  newdate, 02jan1991 to 30dec2013
                delta:  1 day
. * ARCH constraints
. constraint 1 [L.ARCH]1_1 = [L.ARCH]2_1
. constraint 2 [L.ARCH]1_1 = [L.ARCH]2_2
. constraint 3 [L.ARCH]1_1 = [L.ARCH]3_1
. constraint 4 [L.ARCH]1_1 = [L.ARCH]3_2
. constraint 5 [L.ARCH]1_1 = [L.ARCH]3_3
. constraint 6 [L.ARCH]1_1 = [L.ARCH]4_1
. constraint 7 [L.ARCH]1_1 = [L.ARCH]4_2
. constraint 8 [L.ARCH]1_1 = [L.ARCH]4_3
. constraint 9 [L.ARCH]1_1 = [L.ARCH]4_4

. * GARCH constraints
. constraint 10 [L.GARCH]1_1 = [L.GARCH]2_1
. constraint 11 [L.GARCH]1_1 = [L.GARCH]2_2
. constraint 12 [L.GARCH]1_1 = [L.GARCH]3_1
. constraint 13 [L.GARCH]1_1 = [L.GARCH]3_2
. constraint 14 [L.GARCH]1_1 = [L.GARCH]3_3
. constraint 15 [L.GARCH]1_1 = [L.GARCH]4_1
. constraint 16 [L.GARCH]1_1 = [L.GARCH]4_2
. constraint 17 [L.GARCH]1_1 = [L.GARCH]4_3
. constraint 18 [L.GARCH]1_1 = [L.GARCH]4_4
```

We are going to consider one lag for the ARCH part and one for the GARCH part.

```
. mgarch dvech (Dax30 = L.Dax30) (FTSE100 = L.FTSE100) (Nikkei225 = L.Nikkei225)
> (SP500 = L.SP500), arch(1) garch(1)
> constraints(1 2 3 4 5 6 7 8 9 10 11 12 13 14 15 16 17 1 8) nolog
Diagonal vech MGARCH model

Sample: 03jan1991 - 30dec2013                Number of obs   =      5,827
Distribution: Gaussian                       Wald chi2(4)    =     279.29
Log likelihood =  75538.28                   Prob > chi2     =     0.0000
 ( 1)   [L.ARCH]1_1 - [L.ARCH]2_1 = 0
 ( 2)   [L.ARCH]1_1 - [L.ARCH]2_2 = 0
 ( 3)   [L.ARCH]1_1 - [L.ARCH]3_1 = 0
 ( 4)   [L.ARCH]1_1 - [L.ARCH]3_2 = 0
 ( 5)   [L.ARCH]1_1 - [L.ARCH]3_3 = 0
 ( 6)   [L.ARCH]1_1 - [L.ARCH]4_1 = 0
 ( 7)   [L.ARCH]1_1 - [L.ARCH]4_2 = 0
 ( 8)   [L.ARCH]1_1 - [L.ARCH]4_3 = 0
 ( 9)   [L.ARCH]1_1 - [L.ARCH]4_4 = 0
```

```
(10)  [L.GARCH]1_1 - [L.GARCH]2_1 = 0
(11)  [L.GARCH]1_1 - [L.GARCH]2_2 = 0
(12)  [L.GARCH]1_1 - [L.GARCH]3_1 = 0
(13)  [L.GARCH]1_1 - [L.GARCH]3_2 = 0
(14)  [L.GARCH]1_1 - [L.GARCH]3_3 = 0
(15)  [L.GARCH]1_1 - [L.GARCH]4_1 = 0
(16)  [L.GARCH]1_1 - [L.GARCH]4_2 = 0
(17)  [L.GARCH]1_1 - [L.GARCH]4_3 = 0
(18)  [L.GARCH]1_1 - [L.GARCH]4_4 = 0
```

	Coef.	Std. Err.	z	P>|z|	[95% Conf. Interval]	
Dax30						
Dax30						
L1.	-.0881775	.0104573	-8.43	0.000	-.1086734	-.0676816
_cons	.0006998	.0001331	5.26	0.000	.000439	.0009607
FTSE100						
FTSE100						
L1.	-.1074941	.0104446	-10.29	0.000	-.1279652	-.0870231
_cons	.0004173	.0001056	3.95	0.000	.0002103	.0006243
Nikkei225						
Nikkei225						
L1.	-.005904	.0127833	-0.46	0.644	-.0309588	.0191508
_cons	.0001631	.0001617	1.01	0.313	-.0001538	.0004799
SP500						
SP500						
L1.	-.1954765	.0130076	-15.03	0.000	-.220971	-.1699819
_cons	.0006303	.0001042	6.05	0.000	.0004261	.0008345
Sigma0						
1_1	1.60e-06	1.54e-07	10.36	0.000	1.30e-06	1.90e-06
2_1	9.23e-07	9.85e-08	9.37	0.000	7.30e-07	1.12e-06
3_1	6.09e-07	1.06e-07	5.73	0.000	4.01e-07	8.18e-07
4_1	5.59e-07	7.60e-08	7.36	0.000	4.11e-07	7.08e-07
2_2	9.63e-07	1.00e-07	9.62	0.000	7.67e-07	1.16e-06
3_2	4.31e-07	8.37e-08	5.15	0.000	2.67e-07	5.95e-07
4_2	4.15e-07	6.04e-08	6.86	0.000	2.96e-07	5.33e-07
3_3	2.52e-06	2.43e-07	10.40	0.000	2.05e-06	3.00e-06
4_3	2.69e-07	7.80e-08	3.44	0.001	1.16e-07	4.22e-07
4_4	9.30e-07	9.56e-08	9.73	0.000	7.43e-07	1.12e-06
L.ARCH						
1_1	.0471373	.0020519	22.97	0.000	.0431156	.051159
2_1	.0471373	.0020519	22.97	0.000	.0431156	.051159
3_1	.0471373	.0020519	22.97	0.000	.0431156	.051159
4_1	.0471373	.0020519	22.97	0.000	.0431156	.051159
2_2	.0471373	.0020519	22.97	0.000	.0431156	.051159
3_2	.0471373	.0020519	22.97	0.000	.0431156	.051159
4_2	.0471373	.0020519	22.97	0.000	.0431156	.051159
3_3	.0471373	.0020519	22.97	0.000	.0431156	.051159
4_3	.0471373	.0020519	22.97	0.000	.0431156	.051159
4_4	.0471373	.0020519	22.97	0.000	.0431156	.051159

L.GARCH						
1_1	.9451202	.0023507	402.05	0.000	.9405129	.9497276
2_1	.9451202	.0023507	402.05	0.000	.9405129	.9497276
3_1	.9451202	.0023507	402.05	0.000	.9405129	.9497276
4_1	.9451202	.0023507	402.05	0.000	.9405129	.9497276
2_2	.9451202	.0023507	402.05	0.000	.9405129	.9497276
3_2	.9451202	.0023507	402.05	0.000	.9405129	.9497276
4_2	.9451202	.0023507	402.05	0.000	.9405129	.9497276
3_3	.9451202	.0023507	402.05	0.000	.9405129	.9497276
4_3	.9451202	.0023507	402.05	0.000	.9405129	.9497276
4_4	.9451202	.0023507	402.05	0.000	.9405129	.9497276

```
. estimates store scalardvech
```

We first indicate that we want to fit a dvech model by specifying `dvech` just after the `mgarch` command. We then indicate that for each time series, we want to fit an autoregressive model, where each index is a function of just its own lagged values. An interesting feature is the possibility of specifying alternative models for each time series considered. For instance, we could decide to model the level of time series with a VAR model with one term, using a single statement like this:

```
(y1 y2 y3=l.y1 l.y2 l.y3)
```

Or, we may be looking for higher flexibility and thus be interested in specifying alternative models for all the time series considered. For example, for the first time series `y1`, we can adopt an AR(1) model, while for the second time series `y2`, we could choose an AR(2) model. In that case, we should use the following syntax:

```
(y1=l.y1) (y2=l1.y2 l2.y2)
```

We also need to specify the number of lags we want to include by using the `arch()` and `garch()` options. In our case, we add one lag for the ARCH part and one lag for the GARCH part. Finally, we set the constraints we have just created so that we are effectively going to fit a scalar dvech model.

Once the estimation procedure is completed, we are provided with some general information on the fit model, such as the length of the sample size, the distribution under which the maximization was carried out, the final log likelihood, where the maximization algorithm stopped, and a Wald test against the null hypothesis that all the coefficients are equal to 0. In our example, the time series length is 5,827 days (the first day is used with initialization purposes) and the Wald test is highly significant, indicating that our model can describe the time structure of our data in a satisfactory way.

Finally, the table with the output reports the point estimates and standard errors and tests for the significance of the estimated parameters with confidence intervals. We classify the estimated parameters into two main blocks. The first block is regarding the mean equation, and the second block is about the MGARCH model, namely, the intercept Ω (`Sigma0`), the matrix A of lagged standardized errors (`L.ARCH`), and the matrix B

of lagged conditional covariance (`L.GARCH`). We note that, as requested, all the ARCH coefficients are equal to 0.05 and all the GARCH coefficients are equal to 0.95.

We can now fit a proper dvech model using as starting values those obtained from the scalar dvech model.

```
. matrix scalarparam = e(b)

. mgarch dvech (Dax30 = L.Dax30) (FTSE100 = L.FTSE100) (Nikkei225 = L.Nikkei225)
> (SP500 = L.SP500), arch(1) garch(1) from(scalarparam) nolog

Diagonal vech MGARCH model

Sample: 03jan1991 - 30dec2013                    Number of obs    =      5,827
Distribution: Gaussian                           Wald chi2(4)     =     255.06
Log likelihood =  75898.87                       Prob > chi2      =     0.0000
```

	Coef.	Std. Err.	z	P>|z|	[95% Conf.	Interval]
Dax30						
Dax30						
L1.	-.0864081	.0105227	-8.21	0.000	-.1070321	-.065784
_cons	.000953	.0001324	7.20	0.000	.0006936	.0012125
FTSE100						
FTSE100						
L1.	-.0967432	.0105083	-9.21	0.000	-.1173391	-.0761474
_cons	.0005972	.0001046	5.71	0.000	.0003922	.0008022
Nikkei225						
Nikkei225						
L1.	-.0123196	.0133488	-0.92	0.356	-.0384827	.0138435
_cons	.0004371	.0001595	2.74	0.006	.0001245	.0007497
SP500						
SP500						
L1.	-.1893403	.0130676	-14.49	0.000	-.2149524	-.1637282
_cons	.0007632	.0001028	7.42	0.000	.0005618	.0009647
Sigma0						
1_1	1.97e-06	2.43e-07	8.11	0.000	1.50e-06	2.45e-06
2_1	1.07e-06	1.37e-07	7.86	0.000	8.06e-07	1.34e-06
3_1	1.14e-06	3.01e-07	3.78	0.000	5.49e-07	1.73e-06
4_1	6.10e-07	8.14e-08	7.49	0.000	4.50e-07	7.69e-07
2_2	1.06e-06	1.55e-07	6.86	0.000	7.60e-07	1.37e-06
3_2	8.85e-07	2.14e-07	4.14	0.000	4.66e-07	1.30e-06
4_2	4.63e-07	6.56e-08	7.06	0.000	3.35e-07	5.92e-07
3_3	4.71e-06	6.86e-07	6.86	0.000	3.36e-06	6.05e-06
4_3	1.23e-07	6.47e-08	1.90	0.057	-3.61e-09	2.50e-07
4_4	1.14e-06	1.65e-07	6.89	0.000	8.14e-07	1.46e-06
L.ARCH						
1_1	.0539048	.0037186	14.50	0.000	.0466165	.061193
2_1	.0463987	.0031049	14.94	0.000	.0403131	.0524842
3_1	.0192904	.0036408	5.30	0.000	.0121545	.0264263
4_1	.0348038	.0023151	15.03	0.000	.0302664	.0393413

2_2	.0525265	.0038728	13.56	0.000	.044936	.060117
3_2	.0168358	.0033064	5.09	0.000	.0103553	.0233163
4_2	.0304487	.0024049	12.66	0.000	.0257351	.0351622
3_3	.07906	.0063582	12.43	0.000	.0665981	.0915219
4_3	.003091	.0019429	1.59	0.112	-.0007171	.0068991
4_4	.0557718	.0042168	13.23	0.000	.047507	.0640366
L.GARCH						
1_1	.9341379	.0044854	208.26	0.000	.9253466	.9429293
2_1	.9418537	.0038818	242.63	0.000	.9342455	.9494619
3_1	.9507836	.0099278	95.77	0.000	.9313255	.9702417
4_1	.9526631	.0027792	342.79	0.000	.947216	.9581101
2_2	.9370948	.0046561	201.26	0.000	.927969	.9462206
3_2	.9525734	.0091066	104.60	0.000	.9347247	.970422
4_2	.9565671	.0029796	321.03	0.000	.9507271	.9624071
3_3	.8990617	.0077993	115.27	0.000	.8837754	.914348
4_3	.9890951	.0052532	188.28	0.000	.978799	.9993912
4_4	.9322996	.0050408	184.95	0.000	.9224198	.9421795

```
. estimates store dvech
```

Focusing our attention on the ARCH and GARCH parts, we can see that now the parameters are all different among them and statistically significant, suggesting that perhaps the restrictions imposed by the scalar dvech model were too strong. The log likelihood from the unrestricted model is equal to 75,898.87, while the restricted model stopped at 75,538.28. We can use the information criteria to compare the two models.

```
. estimates stats scalardvech dvech
```
Akaike's information criterion and Bayesian information criterion

Model	Obs	ll(null)	ll(model)	df	AIC	BIC
scalardvech	5,827	.	75538.28	20	-151036.6	-150903.2
dvech	5,827	.	75898.87	38	-151721.7	-151468.3

Note: N=Obs used in calculating BIC; see [R] BIC note.

Both the AIC and the BIC clearly indicate that the dvech model should be preferred to the scalar dvech.

To obtain forecasts of all the variances and covariances, we use the **predict** command. We indicate that we want to store all quantities in a set of new variables using a common initial letter, H, and the * indicates that we want to get forecasts about all the quantities. We compute the correlation by using the **generate** command followed by the simple Pearson's correlation coefficient formula. Finally, we plot each time series by using the **tsline** command, and we represent them all in one figure by using the **graph combine** command.

We now represent variances (see figure 4.3) and correlations (see figure 4.4) obtained from the dvech model.

```
. predict H_dvech*, variance
(1 missing value generated)
. tsset date
        time variable:  date, 1/2/1991 to 12/30/2013, but with gaps
                delta:  1 day
. tsline H_dvech_Dax30_Dax30, name(H_dvech_VolDax30)
. tsline H_dvech_FTSE100_FTSE100, name(H_dvech_VolFTSE100)
. tsline H_dvech_Nikkei225_Nikkei225, name(H_dvech_VolNikkei225)
. tsline H_dvech_SP500_SP500, name(H_dvech_VolSP500)
. graph combine H_dvech_VolDax30 H_dvech_VolFTSE100 H_dvech_VolNikkei225
> H_dvech_VolSP500
```

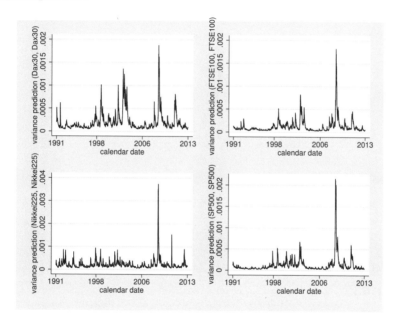

Figure 4.3. Variances from dvech model

```
. generate corr_dvech_daxsp = H_dvech_SP500_Dax30/((H_dvech_Dax30_Dax30)^
> 0.5*(H_dvech_SP500_SP500)^0.5)
(1 missing value generated)
. generate corr_dvech_ftsesp = H_dvech_SP500_FTSE100/((H_dvech_FTSE100_FTSE100)^
> 0.5*(H_dvech_SP500_SP500)^0.5)
(1 missing value generated)
. generate corr_dvech_nikkeisp =
> H_dvech_SP500_Nikkei225/((H_dvech_Nikkei225_Nikkei225)^
> 0.5*(H_dvech_SP500_SP500)^0.5)
(1 missing value generated)
. tsline corr_dvech_daxsp, name(corr_dvech_daxsp)
. tsline corr_dvech_ftsesp, name(corr_dvech_ftsesp)
. tsline corr_dvech_nikkeisp, name(corr_dvech_nikkeisp)
```

```
. graph combine corr_dvech_daxsp corr_dvech_ftsesp corr_dvech_nikkeisp
```

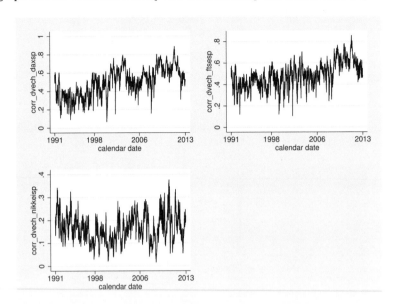

Figure 4.4. Correlations from dvech model

4.4   Nonlinear combination of univariate GARCH—common features

In this section, we report models that may be seen as nonlinear combinations of univariate GARCH models. The nonlinearity depends on the possibility of disentangling the estimation of individual conditional variances from the measure of dependence among the time series, for instance, the conditional correlation matrix or the copula of the conditional joint density. More on this when we present the models in detail.

We now focus on two important advantages of models belonging to this class versus the vech, dvech, and BEKK models presented above. First, when working with a nonlinear combination of univariate GARCH, we can choose an ad-hoc GARCH specification for each historical time series, moving from standard GARCH to threshold or asymmetric GARCH. This feature allows for a high degree of flexibility because we can choose the model that best fits each time series, recognizing the peculiarities of each of them.

The second clear advantage of this kind of model is that it requires a lower number of parameters to be estimated compared with the more traditional multivariate GARCH, especially the vech and BEKK models. The result is an easier maximization problem, even when working with a large number of time series, and we simultaneously gain a better model understanding. Also, the conditional covariance matrix we obtain is positive semidefinite by construction (that means without imposing any constraints), making the estimation procedure simpler from a mathematical point of view (a feature that characterizes even the BEKK model).

We are now going to introduce this class of models by presenting the CCC and the two versions of the DCC model, one of which is by Engle (2002) and the other by Tse and Tsui (2002).

4.4.1 Constant conditional correlation (CCC) GARCH

The main advantage of these models lies in the decomposition of the variance matrix into two parts, conditional volatilities and conditional correlations. For instance, in his CCC model, Bollerslev (1990) specifies the conditional variance–covariance matrix H_t as

$$H_t = D_t R D_t = \left(\rho_{ij} \sqrt{h_{ii,t} h_{jj,t}} \right) \tag{4.7}$$

where D_t is the diagonal matrix of conditional standard deviations computed by univariate GARCH models taking the form

$$
\begin{pmatrix}
h_{11,t}^{1/2} & 0 & \dots & 0 \\
0 & h_{22,t}^{1/2} & \dots & 0 \\
\dots & \dots & \dots & \dots \\
0 & 0 & \dots & h_{mm,t}^{1/2}
\end{pmatrix}
$$

and R is the correlation matrix of innovations ε_t (supposed to be constant for every time t) that we can represent as

$$
\begin{pmatrix}
1 & \rho_{12} & \dots & \rho_{1m} \\
\rho_{21} & 1 & \dots & \rho_{2m} \\
\dots & \dots & \dots & \dots \\
\rho_{m1} & \rho_{m2} & \dots & 1
\end{pmatrix}
$$

The estimation procedure for the CCC model basically consists of two steps. In the first step, we fit m univariate GARCH models to obtain conditional standard deviations $h_{ii,t}$ for all the historical series to populate matrix D_t. In the second step, we estimate the time-invariant correlation matrix R by adopting the sample correlation estimator in (4.8):

$$\bar{p}_{ij} = \frac{\displaystyle\sum_{t=1}^{T} \frac{\varepsilon_{i,t}}{\sqrt{\hat{h}_{i,t}}} \frac{\varepsilon_{j,t}}{\sqrt{\hat{h}_{j,t}}}}{\sqrt{\displaystyle\sum_{t=1}^{T} \left(\frac{\varepsilon_{i,t}}{\sqrt{\hat{h}_{i,t}}}\right)^2 \sum_{t=1}^{T} \left(\frac{\varepsilon_{j,t}}{\sqrt{\hat{h}_{j,t}}}\right)^2}} \tag{4.8}$$

When working with m time series and considering p and q lags, the number of estimates for the CCC model are $(p+q+1)m+\{m(m-1)/2\}$. Therefore, when working with five time series ($m = 5$) and setting $p = q = 1$, we come up with 30 parameters. This is definitely a lower number of parameters compared with the vech, dvech, and BEKK models, where we showed that identical values of m, p, and q gave us 465, 45, and 65 parameters, respectively.

Moreover, the estimation of the CCC model is even simpler than the models considered above because it can be carried out in two steps: in the first step, we fit the five univariate models, each of them involving 3 parameters; in the second step, we deal with the multivariate part, involving 10 parameters for the correlation matrix R. A clear advantage of carrying out the estimation in two steps is that the computational time decreases substantially while keeping estimates consistent.

When working under the assumption of multivariate normal distribution, the log-likelihood function takes the following form:

$$\text{LogL} = -0.5Tm \log(2\pi) - 0.5T \log\{\det(R)\} - \sum_{t=1}^{T} \left[\log\left\{ \det\left(D_t^{1/2}\right) \right\} + 0.5\zeta_t R^{-1} \zeta_t' \right] \tag{4.9}$$

where $\zeta_t = D_t^{-1/2}\varepsilon_t$ is a vector of standardized residuals.

When assuming that errors follow a multivariate Student's t distribution with v degrees of freedom, the log likelihood looks like this:

$$\text{LogL} = T \log \Gamma\left(\frac{v+m}{2}\right) - T \log \Gamma\left(\frac{v}{2}\right) - T\frac{m}{2} \log\{(v-2)\pi\} - 0.5T \log\{\det(R)\} - $$

$$\sum_{t=1}^{T} \left[\log\left\{ \det\left(D_t^{1/2}\right) \right\} + \frac{v+m}{2} \log\left(1 + \frac{\zeta_t R^{-1} \zeta_t'}{v-2}\right) \right] \tag{4.10}$$

By default, the estimation is carried out by ML under the assumption of multivariate normal distribution, but it is possible to select the multivariate Student's t. We can choose the multivariate distribution under which to carry out the estimation by using the `distribution()` option. When making this decision, we have to bear in mind that when we are adopting the multivariate normal distribution, we will find ourselves in one of these situations:

1. If errors are normally distributed and the estimation is carried out by ML or QML, then estimates are consistent and efficient.

2. If errors are not normally distributed, then estimates obtained by QML will be consistent but not efficient.

When opting for the multivariate Student's t distribution, we can find ourselves in one of these two cases:

1. Errors effectively come from a multivariate Student's t, and therefore estimates obtained by ML are consistent and efficient.

2. Errors do not follow a multivariate Student's t, and estimates obtained by QML will not be consistent.

Empirical application

The CCC, one of the MGARCH models implemented in Stata, is fit by the `mgarch ccc` command. As already said, the first step is deciding the formulation for the univariate volatilities. You can specify distinct covariates and distinct ARCH and GARCH terms for each series. We are not allowed to choose among the alternative GARCH specifications, for instance, an asymmetric or a nonlinear GARCH. We can decide to augment the terms of the univariate models by adding some extra variables via the `het()` option, exactly as already described for the univariate GARCH model.

For instance, considering the same four time series presented above (S&P 500, FTSE 100, DAX 30, and Nikkei 225), we start by fitting a CCC model where we assume a VAR model with one term for the levels and a GARCH(1,1) model for the univariate volatilities.

```
. tsset newdate
        time variable:  newdate, 02jan1991 to 30dec2013
                delta:  1 day
. mgarch ccc
> (Dax30 FTSE100 Nikkei225 SP500=L.Dax30 L.FTSE100 L.Nikkei225 L.SP500),
> arch(1) garch(1) nolog
Constant conditional correlation MGARCH model
```

Sample: 03jan1991 - 30dec2013   Number of obs   =   5,827
Distribution: Gaussian   Wald chi2(16)   =   2132.43
Log likelihood = 76237.47   Prob > chi2   =   0.0000

	Coef.	Std. Err.	z	P>|z|	[95% Conf. Interval]	
Dax30						
Dax30						
L1.	-.1218862	.0193849	-6.29	0.000	-.15988	-.0838925
FTSE100						
L1.	-.0617681	.0235022	-2.63	0.009	-.1078315	-.0157047

Nikkei225						
L1.	-.0104709	.0108637	-0.96	0.335	-.0317634	.0108216
SP500						
L1.	.3956999	.0183696	21.54	0.000	.3596962	.4317036
_cons	.0008853	.0001358	6.52	0.000	.0006191	.0011515
ARCH_Dax30						
arch						
L1.	.0622677	.0053094	11.73	0.000	.0518616	.0726738
garch						
L1.	.9085628	.0076539	118.71	0.000	.8935613	.9235642
_cons	4.08e-06	4.78e-07	8.54	0.000	3.14e-06	5.02e-06
FTSE100						
Dax30						
L1.	-.0555665	.0141773	-3.92	0.000	-.0833535	-.0277795
FTSE100						
L1.	-.1214051	.0187131	-6.49	0.000	-.1580821	-.0847281
Nikkei225						
L1.	-.026221	.0085288	-3.07	0.002	-.0429371	-.0095048
SP500						
L1.	.3335138	.0144318	23.11	0.000	.3052279	.3617997
_cons	.0005449	.0001044	5.22	0.000	.0003403	.0007495
ARCH_FTSE100						
arch						
L1.	.0686346	.0057296	11.98	0.000	.0574048	.0798643
garch						
L1.	.9033866	.0078755	114.71	0.000	.8879508	.9188223
_cons	2.34e-06	2.97e-07	7.91	0.000	1.76e-06	2.93e-06
Nikkei225						
Dax30						
L1.	.0739991	.0179	4.13	0.000	.0389159	.1090824
FTSE100						
L1.	.140731	.022908	6.14	0.000	.0958322	.1856299
Nikkei225						
L1.	-.0964351	.0129252	-7.46	0.000	-.1217681	-.0711021
SP500						
L1.	.4016654	.0177488	22.63	0.000	.3668784	.4364524
_cons	.0001309	.0001436	0.91	0.362	-.0001506	.0004123

ARCH_Nikk~225						
arch						
L1.	.0901925	.0074839	12.05	0.000	.0755243	.1048607
garch						
L1.	.8921382	.0085044	104.90	0.000	.8754698	.9088066
_cons	3.52e-06	5.75e-07	6.13	0.000	2.40e-06	4.65e-06
SP500						
Dax30						
L1.	.0276987	.013916	1.99	0.047	.0004238	.0549737
FTSE100						
L1.	-.0259256	.0179709	-1.44	0.149	-.0611479	.0092967
Nikkei225						
L1.	-.0083504	.0082555	-1.01	0.312	-.0245309	.0078302
SP500						
L1.	-.0418941	.0157242	-2.66	0.008	-.072713	-.0110753
_cons	.0007233	.0001047	6.91	0.000	.0005181	.0009284
ARCH_SP500						
arch						
L1.	.0718015	.0059223	12.12	0.000	.060194	.0834089
garch						
L1.	.9084066	.0072323	125.60	0.000	.8942315	.9225816
_cons	1.85e-06	2.43e-07	7.61	0.000	1.37e-06	2.33e-06
corr(Dax30, FTSE100)	.6900719	.0069469	99.34	0.000	.6764563	.7036876
corr(Dax30, Nikkei225)	.2189473	.0124957	17.52	0.000	.1944562	.2434384
corr(Dax30, SP500)	.5030651	.0099018	50.81	0.000	.4836578	.5224723
corr(FTSE100, Nikkei225)	.2130825	.0125313	17.00	0.000	.1885216	.2376435
corr(FTSE100, SP500)	.4978592	.0099055	50.26	0.000	.4784448	.5172735
corr(Nik~225, SP500)	.1396129	.0128631	10.85	0.000	.1144016	.1648242

To fit a VAR model, we specified (Dax30 FTSE100 Nikkei225 SP500=L.Dax30 L.FTSE100 L.Nikkei225 L.SP500), indicating that each time series is modeled as a function of the lagged values of all four indexes.

Estimation of a CCC model is simpler than estimation of a dvech model, because it involves a lower number of parameters and because we do not need to add any constraints to ensure that the covariance matrix is positive semidefinite. The final table is organized as follows. For each time series, we can see the parameters regarding the mean equation and the univariate volatilities, and the correlations are reported in the lower part of the table. Almost all the coefficients are statistically significant; two exceptions are the mean equation of the S&P 500, where the lagged coefficients of FTSE 100 and Nikkei 225 are not statistically significant, and the mean equation of DAX 30, where the lagged value of Nikkei 225 is not statistically significant.

We could even decide to specify an alternative set of covariates for each of the time series considered. For instance, for the model just fit above, we could decide to keep for each stock index only the covariates that were statistically significant. Therefore, we fit the following model:

```
mgarch ccc (Dax30 = L.Dax30 L.FTSE100 L.SP500)                ///
      (FTSE100 = L.Dax30 L.FTSE100 L.Nikkei225 L.SP500)       ///
      (Nikkei225 = L.Dax30 L.FTSE100 L.Nikkei225 L.SP500)     ///
      (SP500 = L.Dax30 L.SP500), arch(1) garch(1) nolog
```

Moreover, using the equation options `arch()` and `garch()`, we can specify alternative GARCH models for the univariate time series considered, as shown in the command below.

```
. mgarch ccc (Dax30 = L.Dax30 L.FTSE100 L.SP500, arch(1) garch(1))
> (FTSE100 = L.Dax30 L.FTSE100 L.Nikkei225 L.SP500, arch(1/2) garch(1))
> (Nikkei225 = L.Dax30 L.FTSE100 L.Nikkei225 L.SP500, arch(1) garch(1/2))
> (SP500 = L.Dax30 L.SP500, arch(1) garch(1)), nolog
Constant conditional correlation MGARCH model
```

Sample: 03jan1991 - 30dec2013   Number of obs   =   5,827
Distribution: Gaussian   Wald chi2(13)   =   2119.36
Log likelihood =   76235.7   Prob > chi2   =   0.0000

	Coef.	Std. Err.	z	P>|z|	[95% Conf. Interval]	
Dax30						
Dax30						
L1.	-.1328055	.0183607	-7.23	0.000	-.1687918	-.0968192
FTSE100						
L1.	-.0469546	.0202552	-2.32	0.020	-.0866541	-.007255
SP500						
L1.	.3933379	.0182786	21.52	0.000	.3575125	.4291632
_cons	.0008832	.0001358	6.50	0.000	.0006169	.0011494
ARCH_Dax30						
arch						
L1.	.0622349	.005312	11.72	0.000	.0518236	.0726462
garch						
L1.	.9086134	.0076641	118.55	0.000	.8935921	.9236348
_cons	4.08e-06	4.79e-07	8.52	0.000	3.14e-06	5.02e-06
FTSE100						
Dax30						
L1.	-.0635188	.0134542	-4.72	0.000	-.0898885	-.0371491
FTSE100						
L1.	-.1091353	.0163278	-6.68	0.000	-.1411372	-.0771333

Nikkei225						
L1.	-.0198419	.0062016	-3.20	0.001	-.0319969	-.007687
SP500						
L1.	.331569	.014368	23.08	0.000	.3034082	.3597298
_cons	.0005426	.0001044	5.20	0.000	.000338	.0007472
ARCH_FTSE100						
arch						
L1.	.068319	.0117242	5.83	0.000	.0453401	.0912979
L2.	.0004193	.0129686	0.03	0.974	-.0249986	.0258372
garch						
L1.	.9031644	.0088947	101.54	0.000	.8857312	.9205977
_cons	2.35e-06	3.18e-07	7.40	0.000	1.73e-06	2.98e-06
Nikkei225						
Dax30						
L1.	.07156	.0178428	4.01	0.000	.0365888	.1065311
FTSE100						
L1.	.1449393	.0227172	6.38	0.000	.1004143	.1894643
Nikkei225						
L1.	-.0939946	.0127337	-7.38	0.000	-.1189522	-.0690371
SP500						
L1.	.401424	.0177577	22.61	0.000	.3666196	.4362283
_cons	.0001282	.0001436	0.89	0.372	-.0001533	.0004097
ARCH_Nikk~225						
arch						
L1.	.0949207	.0121639	7.80	0.000	.07108	.1187614
garch						
L1.	.8222253	.1374503	5.98	0.000	.5528277	1.091623
L2.	.0643232	.1262499	0.51	0.610	-.1831221	.3117684
_cons	3.69e-06	6.81e-07	5.41	0.000	2.35e-06	5.02e-06
SP500						
Dax30						
L1.	.012922	.0108602	1.19	0.234	-.0083637	.0342076
SP500						
L1.	-.0466764	.0153928	-3.03	0.002	-.0768457	-.0165071
_cons	.0007215	.0001046	6.89	0.000	.0005164	.0009266
ARCH_SP500						
arch						
L1.	.0718075	.0059383	12.09	0.000	.0601686	.0834463
garch						
L1.	.9082991	.0072643	125.04	0.000	.8940614	.9225368

_cons	1.86e-06	2.45e-07	7.60	0.000	1.38e-06	2.34e-06
corr(Dax30, FTSE100)	.6900476	.0069438	99.38	0.000	.676438	.7036571
corr(Dax30, Nikkei225)	.2187894	.0124977	17.51	0.000	.1942944	.2432843
corr(Dax30, SP500)	.5029008	.0098958	50.82	0.000	.4835053	.5222963
corr(FTSE100, Nikkei225)	.2129415	.0125328	16.99	0.000	.1883776	.2375054
corr(FTSE100, SP500)	.4975947	.0099052	50.24	0.000	.4781808	.5170086
corr(Nik~225, SP500)	.1394817	.0128634	10.84	0.000	.11427	.1646934

```
. estimates store CCC
```

Here we specify a GARCH(1,1) for the DAX 30 and the S&P 500, while we select a GARCH(1,2) for the Nikkei 225 and a GARCH(2,1) for the FTSE 100.

All the correlation coefficients reported in the output above are positive and statistically different from 0, indicating that stock indexes tend to move in a similar way. The highest estimated correlation is between DAX 30 and FTSE 100 (0.69), followed by the one between DAX 30 and S&P 500 (0.50) and between FTSE 100 and S&P 500 (0.50). The lowest correlations correspond to the Nikkei 225, ranging from a minimum of 0.14 when evaluated with the S&P 500 to a maximum of 0.22 when evaluated with the DAX 30.

We now present the variances from the CCC model just reported above:

```
. predict H*, variance
. tsset date
        time variable:  date, 1/2/1991 to 12/30/2013, but with gaps
                delta:  1 day
. tsline H_Dax30_Dax30, name(VolDax30)
. tsline H_FTSE100_FTSE100, name(VolFTSE100)
. tsline H_Nikkei225_Nikkei225, name(VolNikkei225)
. tsline H_SP500_SP500, name(VolSP500)
. graph combine VolDax30 VolFTSE100 VolNikkei225 VolSP500
```

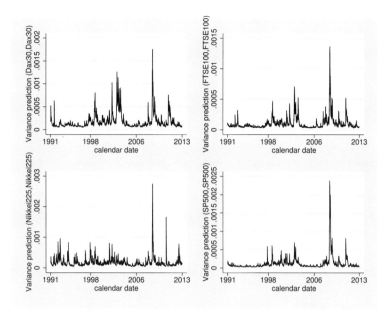

Figure 4.5. Estimated variances from the CCC model

Note in figure 4.5 the strong evidence of heteroskedasticity, which is the presence of periods of systematic low volatility followed by periods of systematic high volatility. In particular, we can see that the sharpest increase of volatility has occurred in correspondence to the alternative crises that financial markets experienced during the time window analyzed. The strongest increase in volatilities occurred during the financial crisis of years 2007–2009, and it affected all four of the indexes considered. We can see a significant increase in variances for the S&P 500, DAX 30, and FTSE 100 during the dot-com bubble of 2001 and the sovereign crisis of years 2010–2012.

We now represent in figure 4.6 the correlations from the CCC model.

```
predict corr*, correlation
tsline corr_FTSE100_Dax30, name(corrftsedax)
tsline corr_Nikkei225_Dax30, name(corrnikkeidax)
tsline corr_SP500_Dax30, name(corrspdax)
tsline corr_Nikkei225_FTSE100, name(corrnikkeiftse)
tsline corr_SP500_FTSE100, name(corrspftse)
tsline corr_SP500_Nikkei225, name(corrsonikkei)
graph combine corrftsedax corrnikkeidax corrspdax corrnikkeiftse corrspftse  ///
    corrsonikkei
```

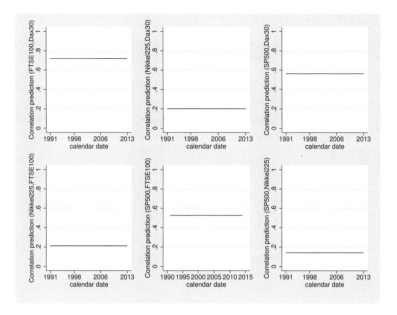

Figure 4.6. Correlations from the CCC model

Figure 4.6 gives a very clear idea about what we mean by constant correlation: the correlation is equal to a straight line throughout the time considered. In particular, the correlations are equal to the values reported in the output of the CCC model. For instance, the correlation between FTSE 100 and DAX 30 is equal to 0.69, which corresponds exactly to the level reported in figure 4.6.

4.4.2   Dynamic conditional correlation (DCC) model

The assumption of constant conditional correlations in the CCC model is quite unrealistic for many empirical applications. To cope with this restrictive assumption and therefore model the dynamic nature of correlations, Engle (2002) and Tse and Tsui (2002) proposed almost contemporaneously a new class of MGARCH model, the DCC. We describe both versions below.

Dynamic conditional correlation Engle (DCCE) model

The DCC model by Engle (2002) relies on the same decomposition of the variance–covariance matrix as the CCC model does, but here correlations are time dependent. Therefore, we can rewrite the CCC decomposition in (4.7) as

$$H_t = D_t R_t D_t \tag{4.11}$$

where the conditional correlation matrix R_t is time dependent. R_t takes the following form:

$$R_t = \text{diag}\left(q_{11,t}^{-1/2}, q_{22,t}^{-1/2}, \ldots, q_{mm,t}^{-1/2}\right) Q_t \text{diag}\left(q_{11,t}^{-1/2}, q_{22,t}^{-1/2}, \ldots, q_{mm,t}^{-1/2}\right) \qquad (4.12)$$

The elements of matrix Q_t follow a univariate GARCH model; for instance, Engle (2002) assumes a standard GARCH(1,1),

$$Q_t = V_{ij} + \lambda_1 \frac{\varepsilon_{t-1}}{\sqrt{h_{t-1}}} \left(\frac{\varepsilon_{t-1}}{\sqrt{h_{t-1}}}\right)' + \lambda_2 Q_{t-1} \qquad (4.13)$$

where V_{ij} is a matrix of long-run or unconditional correlations, and λ_1 and λ_2 are parameters that are supposed to be time invariant. The obtained covariance matrix is positive semidefinite, and the only constraint we have to add to ensure that the process is stationary is $\lambda_1 + \lambda_2 < 1$.

Note that parameters λ_1 and λ_2 are scalars, and they take the same values for all the time series considered. Therefore, we assume that all the time series follow the same dynamic, that is, they are all characterized by the same values of λ_1 and λ_2.

We can introduce an additional simplification in (4.13), substituting V_{ij} with $(1 - \lambda_1 - \lambda_2)\overline{Q}$, where $\overline{Q}$ is the sample counterpart of the unconditional correlation matrix. This approach, called variance targeting, allows us to avoid estimating the $m(m-1)/2$ elements of matrix V_{ij}. By applying variance targeting, we can rewrite (4.13) as

$$Q_t = (1 - \lambda_1 - \lambda_2)\overline{Q} + \lambda_1 \frac{\varepsilon_{t-1}}{\sqrt{h_{t-1}}} \left(\frac{\varepsilon_{t-1}}{\sqrt{h_{t-1}}}\right)' + \lambda_2 Q_{t-1} \qquad (4.14)$$

Regardless of whether variance targeting is applied, we can adopt the two-step estimation procedure even for the DCC model. In the first step, we fit m univariate GARCH models to obtain the conditional standard deviations composing the diagonal elements of matrices D_t in (4.11). At this point, we can compute the standardized residuals as $\zeta_t = D_t^{-1/2}\varepsilon_t$ and proceed by calculating the correlation matrices R_t, by applying either (4.12) or (4.13), where parameters are estimated by maximum likelihood.

Aside from the two-step estimation procedure, we can always proceed in one-step estimation by adopting the full log-likelihood function working on the parameters all together. Therefore, according to the assumption about the multivariate distribution, we could work with a Gaussian log-likelihood function,

$$\text{LogL} = -0.5Tm\log(2\pi) - 0.5T\log\{\det(R_t)\} - \sum_{t=1}^{T}\left[\log\left\{\det\left(D_t^{1/2}\right)\right\} + 0.5\zeta_t R_t^{-1}\zeta_t'\right] \qquad (4.15)$$

or with a Student's t log-likelihood function,

$$\mathrm{LogL} = T \log \Gamma \left(\frac{v+m}{2} \right) - T \log \Gamma \left(\frac{v}{2} \right) - T \frac{m}{2} \log \left\{ (v-2)\,\pi \right\} - 0.5T \log \left\{ \det (R_t) \right\}$$

$$- \sum_{t=1}^{T} \left[\log \left\{ \det \left(D_t^{1/2} \right) \right\} + \frac{v+m}{2} \log \left(1 + \frac{\zeta_t R_t^{-1} \zeta_t'}{v-2} \right) \right] \tag{4.16}$$

Note that (4.15) and (4.16) are identical to their counterparts for the CCC model, (4.9) and (4.10), respectively, with the exception of the conditional correlation matrix that here is time dependent, R_t, but is not in the CCC model.

If we opt for the two-step estimation, we can decompose the log-likelihood function in (4.15) and (4.16) into two parts, with the first covering the univariate part (both mean and variance equations) and the second concerning the dynamic of correlations. An inefficient but consistent estimator of R_t is the identity matrix. Using this assumption, we can write the log-likelihood function as the sum of

$$\mathrm{LogL1}(\theta_1) = -\frac{1}{2} \sum_{t=1}^{T} \left[\log\{\mathrm{diag}(D_t)\} + D_t^{-1} \varepsilon_t^2 \right]$$

where θ_1 is the first set of parameters, comprising all the univariate GARCH parameters. The second step of estimation relies on the following log-likelihood function:

$$\mathrm{LogL2}(\theta_2 | \theta_1) = -\frac{1}{2} \sum_{t=1}^{T} \left\{ \log |R_t| + \left(D_t^{-1} \varepsilon_t \right)' R_t^{-1} \left(D_t^{-1} \varepsilon_t \right) \right\}$$

The maximization of this second log likelihood is carried out conditional on the first set of parameters, θ_1. In this second step, we get estimates for the set of parameters θ_2, comprising parameters governing the time dynamic of correlations, a and b in (4.13) or (4.14), with the coefficient of the intercept matrix (when not working under variance targeting).

Empirical application

We now illustrate the DCCE model with the four stock indexes of our dataset. We start by fitting a DCC model with one lag for both the ARCH and the GARCH parts. For the mean part, we adopt a VAR(1) model.

```
. tsset newdate
        time variable:  newdate, 02jan1991 to 30dec2013
                delta:  1 day
. mgarch dcc (Dax30 FTSE100 Nikkei225 SP500=L.Dax30 L.FTSE100 L.Nikkei225 L.SP500),
> arch(1) garch(1) nolog

Dynamic conditional correlation MGARCH model

Sample: 03jan1991 - 30dec2013                    Number of obs   =      5,827
Distribution: Gaussian                           Wald chi2(16)   =    2144.22
Log likelihood = 76682.15                        Prob > chi2     =     0.0000
```

	Coef.	Std. Err.	z	P>|z|	[95% Conf. Interval]	
Dax30						
Dax30						
L1.	-.1204914	.0176933	-6.81	0.000	-.1551697	-.0858131
FTSE100						
L1.	-.0356701	.0215186	-1.66	0.097	-.0778457	.0065055
Nikkei225						
L1.	-.010655	.0101359	-1.05	0.293	-.0305209	.0092109
SP500						
L1.	.3840385	.0174138	22.05	0.000	.3499082	.4181689
_cons	.0008083	.0001274	6.34	0.000	.0005586	.001058
ARCH_Dax30						
arch						
L1.	.0637442	.0045284	14.08	0.000	.0548686	.0726197
garch						
L1.	.9235316	.0055034	167.81	0.000	.9127451	.934318
_cons	2.48e-06	3.14e-07	7.90	0.000	1.87e-06	3.10e-06
FTSE100						
Dax30						
L1.	-.0709634	.0132324	-5.36	0.000	-.0968984	-.0450285
FTSE100						
L1.	-.0903207	.0176058	-5.13	0.000	-.1248274	-.055814
Nikkei225						
L1.	-.0279059	.0081277	-3.43	0.001	-.0438359	-.011976
SP500						
L1.	.3458872	.014006	24.70	0.000	.3184359	.3733386
_cons	.0004791	.0000999	4.80	0.000	.0002833	.0006749
ARCH_FTSE100						
arch						
L1.	.0680758	.0049976	13.62	0.000	.0582807	.077871
garch						
L1.	.9217678	.00573	160.87	0.000	.9105371	.9329984
_cons	1.31e-06	1.93e-07	6.77	0.000	9.29e-07	1.69e-06
Nikkei225						
Dax30						
L1.	.0770802	.0179239	4.30	0.000	.04195	.1122104
FTSE100						
L1.	.149975	.0229475	6.54	0.000	.1049988	.1949511

Nikkei225						
L1.	-.0981265	.012914	-7.60	0.000	-.1234375	-.0728154
SP500						
L1.	.397336	.0178562	22.25	0.000	.3623384	.4323335
_cons	.000105	.000143	0.73	0.463	-.0001752	.0003852
ARCH_Nikk~225						
arch						
L1.	.094264	.0077506	12.16	0.000	.0790732	.1094549
garch						
L1.	.889886	.0085812	103.70	0.000	.8730671	.9067049
_cons	3.60e-06	5.85e-07	6.15	0.000	2.45e-06	4.74e-06
SP500						
Dax30						
L1.	.0258705	.0131375	1.97	0.049	.0001214	.0516196
FTSE100						
L1.	.0085114	.0171116	0.50	0.619	-.0250268	.0420496
Nikkei225						
L1.	-.014192	.0079346	-1.79	0.074	-.0297435	.0013595
SP500						
L1.	-.0350301	.0152815	-2.29	0.022	-.0649812	-.0050789
_cons	.0006714	.0001005	6.68	0.000	.0004744	.0008684
ARCH_SP500						
arch						
L1.	.0720172	.0057486	12.53	0.000	.0607502	.0832842
garch						
L1.	.9176668	.0063759	143.93	0.000	.9051702	.9301633
_cons	1.29e-06	1.96e-07	6.62	0.000	9.11e-07	1.68e-06
corr(Dax30,						
FTSE100)	.8211129	.0199594	41.14	0.000	.7819933	.8602325
corr(Dax30,						
Nikkei225)	.230072	.0539162	4.27	0.000	.1243981	.3357458
corr(Dax30,						
SP500)	.580286	.037353	15.54	0.000	.5070755	.6534965
corr(FTSE100,						
Nikkei225)	.2222712	.0541689	4.10	0.000	.1161021	.3284403
corr(FTSE100,						
SP500)	.5542778	.0393509	14.09	0.000	.4771514	.6314041
corr(Nik~225,						
SP500)	.1523959	.0553386	2.75	0.006	.0439343	.2608575
Adjustment						
lambda1	.0117339	.0008749	13.41	0.000	.0100191	.0134487
lambda2	.9847641	.0011405	863.41	0.000	.9825287	.9869996

```
. estimates store DCC
```

We can see that the syntax for fitting a DCC model is quite similar to that used to fit a CCC model.

The structure of the DCC output table is also very similar to the one obtained for the CCC model. The upper part of the table reports parameter estimates for the VAR model and for univariate GARCH models, while the next part reports the elements of the long-run matrix V. The novelty here lies in the last part of the output, titled `Adjustment`, which reports parameter estimates of λ_1 and λ_2 of (4.13).

The estimated coefficients of the VAR and univariate GARCH models are close to those obtained in the CCC model, in terms of both magnitude and statistical significance. We note some differences when focusing on the correlation coefficients, which are somewhat higher for the DCC. For instance, for the pair DAX 30 and FTSE 100, we obtain an estimate of 0.69 for the CCC and of 0.82 for the DCC; for the pair FTSE 100 and S&P 500, those values are equal to 0.50 and 0.55, respectively.

The most relevant difference between the CCC and DCC models is the introduction of the two parameters `lambda1` and `lambda2`, which take the values 0.01 and 0.98. These values suggest that correlations are highly persistent, which means that the matrix Q_t is highly dependent on the lagged matrix Q_{t-1}.

We now formally test whether the time-dependent correlations introduced by the DCC allow this model to outperform the CCC. For this purpose, we apply the likelihood-ratio test by using the `lrtest` command:

```
. lrtest CCC DCC
Likelihood-ratio test                          LR chi2(3)  =     892.90
(Assumption: CCC nested in DCC)                Prob > chi2 =     0.0000
```

The likelihood-ratio test allows us to reject the null hypothesis about the equivalence of the two models. We conclude that the dataset does not support the assumption that correlations among stock indexes are constant through time. The result is confirmed by the information criteria.

```
. estimates stats CCC DCC
Akaike´s information criterion and Bayesian information criterion
```

Model	Obs	ll(null)	ll(model)	df	AIC	BIC
CCC	5,827	.	76235.7	37	−152397.4	−152150.6
DCC	5,827	.	76682.15	40	−153284.3	−153017.5

Note: N=Obs used in calculating BIC; see [R] BIC note.

Two ad-hoc tests were proposed to test whether the assumption of constant correlations is too strict and therefore a DCC model is more appropriate than a CCC model. Engle and Sheppard (2001) propose the first one, relying on the following hypotheses:

$$H_0\colon\ R_t = \overline{R}\quad \forall t$$
$$H_1\colon\ \mathrm{vech}(R_t) = \mathrm{vech}(\overline{R}) + \beta_1 \mathrm{vech}\,(R_{t-1}) + \cdots + \beta_q \mathrm{vech}\,(R_{t-q})$$

The null hypothesis is that the conditional correlation matrix R_t is constant throughout the sample and equal to $\overline{R}$, while under the alternative, the correlation matrix follows a specific time dynamic.

Tse (2000) proposes a second test, evaluating the following two hypotheses:

$$H_0\colon\ h_{ij,t} = \rho_{ij}\,\sqrt{h_{ii,t}h_{jj,t}}$$
$$H_1\colon\ h_{ij,t} = \rho_{ij,t}\,\sqrt{h_{ii,t}h_{jj,t}}$$

The null hypothesis is that the correlations are constant through the time considered, p_{ij}, against the alternative that correlations are time dependent, $p_{ij,t}$. The test evaluating these two alternative hypotheses is an LM-type test, and its limiting distribution under H_0 is therefore a $\chi^2_{m(m-1)/2}$.

We now represent variances (see figure 4.7) and correlations (see figure 4.8) obtained from the just fit DCC model.

```
. tsset date
        time variable:  date, 1/2/1991 to 12/30/2013, but with gaps
                delta:  1 day
. predict H_DCC*, variance
. tsline H_DCC_Dax30_Dax30, name(VolDCCDax30)
. tsline H_DCC_FTSE100_FTSE100, name(VolDCCFTSE100)
. tsline H_DCC_Nikkei225_Nikkei225, name(VolDCCNikkei225)
. tsline H_DCC_SP500_SP500, name(VolDCCSP500)
. graph combine VolDCCDax30 VolDCCFTSE100 VolDCCNikkei225 VolDCCSP500
```

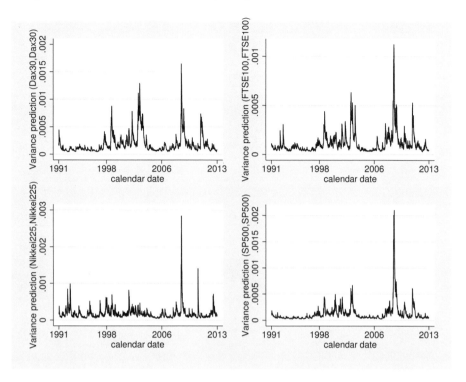

Figure 4.7. Variances from the DCC model

```
. predict corr_DCC*, correlation
. tsline corr_DCC_FTSE100_Dax30, ylabel(0(.2)1) name(corrdccftsedax)
. tsline corr_DCC_Nikkei225_Dax30, ylabel(0(.2)1) name(corrdccnikkeidax)
. tsline corr_DCC_SP500_Dax30, ylabel(0(.2)1) name(corrdccspdax)
. tsline corr_DCC_Nikkei225_FTSE100, ylabel(0(.2)1) name(corrdccnikkeiftse)
. tsline corr_DCC_SP500_FTSE100, ylabel(0(.2)1) name(corrdccspftse)
. tsline corr_DCC_SP500_Nikkei225, ylabel(0(.2)1) name(corrdccspnikkei)
. graph combine corrdccftsedax corrdccnikkeidax corrdccspdax corrdccnikkeiftse
> corrdccspftse corrdccspnikkei
```

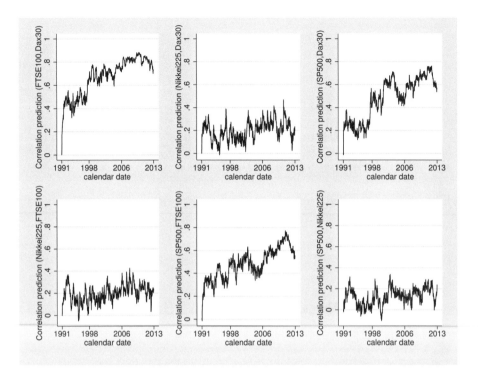

Figure 4.8. Correlations from the DCC model

As already mentioned, the main improvement of the DCC model over the CCC model is that it recognizes the dynamic nature of correlations. We can appreciate the importance of this feature by comparing figure 4.8 with figure 4.6, highlighting the assumption about constant correlations in the CCC model that we remove in the DCC model. In particular, from figure 4.8 we can see that correlations among S&P 500, DAX 30, and FTSE 100 have increased by far in the most recent years, with the pair formed by FTSE 100 and DAX 30 almost reaching the upper bound of 1 during the 2008 financial crisis. Considering the correlations of these three stock indexes with the Nikkei 225, we note that they are lower, although positive, and that they all show a drop corresponding to the burst of the European sovereign crisis, starting in 2010.

We can verify whether the DCC model is correctly specified by evaluating the behavior of its standardized residuals. We first have to compute the standardized residuals, following the procedure described in chapter 3, and then we evaluate whether they still exhibit some time structure by applying the `estat archlm` test.

```
. tsset newdate
        time variable:  newdate, 02jan1991 to 30dec2013
                delta:  1 day
. predict res*, residuals
. predict var*, variance
. generate stdresDax = res_Dax30/var_Dax30_Dax30^0.5
(1 missing value generated)
. generate stdresSP500 = res_SP500/var_SP500_SP500^0.5
(1 missing value generated)
. generate stdresNikkei225 = res_Nikkei225/var_Nikkei225_Nikkei225^0.5
(1 missing value generated)
. generate stdresFtse = res_FTSE100/var_FTSE100_FTSE100 ^0.5
(1 missing value generated)
. regress stdresDax
```

  (*output omitted*)

```
. estat archlm, lags(1,5,10)
LM test for autoregressive conditional heteroskedasticity (ARCH)
```

lags(p)	chi2	df	Prob > chi2
1	0.680	1	0.4095
5	3.491	5	0.6247
10	3.665	10	0.9612

        H0: no ARCH effects   *vs.* H1: ARCH(p) disturbance

```
. regress stdresSP500
```

  (*output omitted*)

```
. estat archlm, lags(1,5,10)
LM test for autoregressive conditional heteroskedasticity (ARCH)
```

lags(p)	chi2	df	Prob > chi2
1	4.529	1	0.0333
5	16.561	5	0.0054
10	19.873	10	0.0305

        H0: no ARCH effects   *vs.* H1: ARCH(p) disturbance

```
. regress stdresNikkei225
```

  (*output omitted*)

```
. estat archlm, lags(1,5,10)
LM test for autoregressive conditional heteroskedasticity (ARCH)
```

lags(p)	chi2	df	Prob > chi2
1	9.212	1	0.0024
5	10.034	5	0.0743
10	13.412	10	0.2015

        H0: no ARCH effects   *vs.* H1: ARCH(p) disturbance

```
. regress stdresFtse
```

  (*output omitted*)

```
. estat archlm, lags(1,5,10)
LM test for autoregressive conditional heteroskedasticity (ARCH)
```

lags(p)	chi2	df	Prob > chi2
1	0.049	1	0.8251
5	16.646	5	0.0052
10	23.288	10	0.0097

H0: no ARCH effects   *vs.*   H1: ARCH(p) disturbance

Analyzing the results, we conclude that the only standardized residuals showing absence of heteroskedasticity are those coming from the DAX 30; for the other three stock indexes, we cannot accept the null hypothesis about the absence of heteroskedasticity, at least at some of the lags considered. In particular, for the S&P 500, we find evidence of heteroskedasticity at all the lags evaluated; for the Nikkei 225, we find evidence just when considering one lag; and for the FTSE 100, we find evidence at the 5th and 10th lags.

To remove the heteroskedasticity, we can think about adding some extra terms in the model. We can choose to insert an extra ARCH term where the test fails at recent lags, to capture the short-lived effect, and we can insert an extra GARCH term when it fails at the 10th lag, to account for a more persistent effect.

```
. mgarch dcc (Dax30 = L.Dax30 L.FTSE100 L.SP500 L.Nikkei225, arch(1) garch(1))
> (FTSE100 = L.Dax30 L.FTSE100 L.SP500 L.Nikkei225, arch(1) garch(1/2))
> (Nikkei225 = L.Dax30 L.FTSE100 L.SP500 L.Nikkei225, arch(1/2) garch(1))
> (SP500=L.Dax30 L.FTSE100 L.SP500 L.Nikkei225, arch(1) garch(1/2)) nolog
Dynamic conditional correlation MGARCH model
```

```
Sample: 03jan1991 - 30dec2013              Number of obs   =      5,827
Distribution: Gaussian                     Wald chi2(16)   =    2135.40
Log likelihood =  76685.71                 Prob > chi2     =     0.0000
```

	Coef.	Std. Err.	z	P>|z|	[95% Conf.	Interval]
Dax30						
Dax30						
L1.	-.1202837	.0176964	-6.80	0.000	-.1549679	-.0855994
FTSE100						
L1.	-.0360292	.0215007	-1.68	0.094	-.0781698	.0061115
SP500						
L1.	.3838627	.0174082	22.05	0.000	.3497433	.4179822
Nikkei225						
L1.	-.0102854	.0101399	-1.01	0.310	-.0301592	.0095884
_cons	.0008075	.0001274	6.34	0.000	.0005578	.0010571
ARCH_Dax30						
arch						
L1.	.0638504	.0045561	14.01	0.000	.0549206	.0727802

garch						
L1.	.923365	.0055392	166.70	0.000	.9125084	.9342216
_cons	2.49e-06	3.15e-07	7.89	0.000	1.87e-06	3.11e-06
FTSE100						
Dax30						
L1.	-.0702905	.0132598	-5.30	0.000	-.0962792	-.0443017
FTSE100						
L1.	-.0907584	.0176537	-5.14	0.000	-.1253589	-.0561578
SP500						
L1.	.3458877	.0140085	24.69	0.000	.3184316	.3733438
Nikkei225						
L1.	-.027447	.0081404	-3.37	0.001	-.0434019	-.0114922
_cons	.0004782	.0000999	4.79	0.000	.0002824	.000674
ARCH_FTSE100						
arch						
L1.	.0777622	.0102227	7.61	0.000	.057726	.0977984
garch						
L1.	.7563828	.1533765	4.93	0.000	.4557704	1.056995
L2.	.1543587	.1436306	1.07	0.283	-.1271522	.4358696
_cons	1.48e-06	2.67e-07	5.57	0.000	9.63e-07	2.01e-06
Nikkei225						
Dax30						
L1.	.0778844	.0179449	4.34	0.000	.042713	.1130559
FTSE100						
L1.	.150375	.0229627	6.55	0.000	.105369	.1953809
SP500						
L1.	.3977174	.0178742	22.25	0.000	.3626847	.4327501
Nikkei225						
L1.	-.0985155	.0130144	-7.57	0.000	-.1240233	-.0730078
_cons	.0001012	.000143	0.71	0.479	-.000179	.0003814
ARCH_Nikk~225						
arch						
L1.	.1050672	.0151624	6.93	0.000	.0753493	.134785
L2.	-.0146002	.0170264	-0.86	0.391	-.0479713	.0187709
garch						
L1.	.8945359	.0097199	92.03	0.000	.8754853	.9135866
_cons	3.40e-06	5.96e-07	5.71	0.000	2.23e-06	4.57e-06
SP500						
Dax30						
L1.	.02533	.01311	1.93	0.053	-.0003651	.0510251

FTSE100						
L1.	.0079786	.0170424	0.47	0.640	−.0254238	.0413811
SP500						
L1.	−.0358915	.0150876	−2.38	0.017	−.0654627	−.0063203
Nikkei225						
L1.	−.0139322	.0079125	−1.76	0.078	−.0294405	.0015761
_cons	.0006722	.0001002	6.71	0.000	.0004758	.0008685
ARCH_SP500						
arch						
L1.	.0555638	.0083271	6.67	0.000	.0392429	.0718846
garch						
L1.	1.205105	.1208798	9.97	0.000	.968185	1.442025
L2.	−.2688087	.1127859	−2.38	0.017	−.489865	−.0477525
_cons	1.02e−06	1.94e−07	5.27	0.000	6.42e−07	1.40e−06
corr(Dax30, FTSE100)	.8212047	.0199808	41.10	0.000	.7820429	.8603664
corr(Dax30, Nikkei225)	.2296897	.0540403	4.25	0.000	.1237727	.3356067
corr(Dax30, SP500)	.5796291	.0374602	15.47	0.000	.5062084	.6530497
corr(FTSE100, Nikkei225)	.2221891	.0542855	4.09	0.000	.1157915	.3285867
corr(FTSE100, SP500)	.5539605	.0394445	14.04	0.000	.4766506	.6312703
corr(Nik~225, SP500)	.1516398	.0554723	2.73	0.006	.0429161	.2603635
Adjustment						
lambda1	.0117099	.000873	13.41	0.000	.0099989	.013421
lambda2	.9848038	.0011369	866.24	0.000	.9825756	.987032

```
. estimates store DCCRefined1
```

In this case, we specify alternative equations for each of the time series considered. We model the mean equations of each index as a VAR with one term, and therefore we have specified the lagged value for each of them. For the variance equation, we have chosen GARCH with alternative terms for each time series. For the DAX 30, we adopt a GARCH(1,1); for the FTSE 100, a GARCH(1,2) (note that `garch(1/2)` indicates that for the GARCH part we are including lags 1 and 2); for the Nikkei 225, a GARCH(2,1) (that is, with two lags for the ARCH part and one for the GARCH part); and finally, for the S&P 500, a GARCH(1,2).

Analyzing results from the so-refined DCC model, we can see that the extra parameters are statistically insignificant with only two exceptions: the second term of the GARCH model added for the FTSE 100 and the second ARCH term for the Nikkei 225.

By comparing this model with the restricted one in terms of goodness of fit, we note that the log likelihood of this DCC refined model is equal to 76,685.71, while it was equal to 76,682.15 in the previous model. This slight improvement in the log likelihood comes at the cost of adding three new parameters. We implement the likelihood-ratio test and the information criteria:

```
. lrtest DCC DCCRefined1
Likelihood-ratio test                              LR chi2(3)  =      7.12
(Assumption: DCC nested in DCCRefined1)            Prob > chi2 =    0.0683
. estimates stats DCC DCCRefined1
Akaike's information criterion and Bayesian information criterion
```

Model	Obs	ll(null)	ll(model)	df	AIC	BIC
DCC	5,827	.	76682.15	40	-153284.3	-153017.5
DCCRefined1	5,827	.	76685.71	43	-153285.4	-152998.6

Note: N=Obs used in calculating BIC; see [R] BIC note.

The likelihood-ratio test indicates that we cannot strongly reject the null hypothesis about the equivalence of the two models, because the improvement in the log likelihood is not big enough to justify the inclusion of three new parameters. The p-value of the test, `Prob > chi2`, is in fact equal to 0.0683, which is greater than the standard criteria of 0.05.

We now remove the second GARCH term for the FTSE 100 and the second ARCH term for the Nikkei 225, and just leave the second GARCH term for the S&P 500.

```
. mgarch dcc (Dax30 = L.Dax30 L.FTSE100 L.SP500 L.Nikkei225,arch(1) garch(1))
> (FTSE100 = L.Dax30 L.FTSE100 L.SP500 L.Nikkei225, arch(1) garch(1))
> (Nikkei225 = L.Dax30 L.FTSE100 L.SP500 L.Nikkei225, arch(1) garch(1))
> (SP500 = L.Dax30 L.FTSE100 L.SP500 L.Nikkei225,arch(1) garch(1/2)), nolog
Dynamic conditional correlation MGARCH model
Sample: 03jan1991 - 30dec2013                      Number of obs   =      5,827
Distribution: Gaussian                             Wald chi2(16)   =    2145.60
Log likelihood =  76684.71                         Prob > chi2     =     0.0000
```

	Coef.	Std. Err.	z	P>|z|	[95% Conf. Interval]	
Dax30						
Dax30						
L1.	-.120632	.0176967	-6.82	0.000	-.1553169	-.0859471
FTSE100						
L1.	-.0358973	.0215125	-1.67	0.095	-.0780611	.0062664
SP500						
L1.	.3839178	.0174147	22.05	0.000	.3497856	.4180501
Nikkei225						
L1.	-.0106667	.0101362	-1.05	0.293	-.0305332	.0091998
_cons	.0008095	.0001274	6.36	0.000	.0005599	.0010592

ARCH_Dax30						
arch						
L1.	.0641811	.004566	14.06	0.000	.0552319	.0731303
garch						
L1.	.9229612	.0055488	166.34	0.000	.9120857	.9338366
_cons	2.51e-06	3.16e-07	7.92	0.000	1.89e-06	3.13e-06
FTSE100						
Dax30						
L1.	-.0710009	.0132322	-5.37	0.000	-.0969354	-.0450663
FTSE100						
L1.	-.0905591	.0176013	-5.15	0.000	-.125057	-.0560612
SP500						
L1.	.3457599	.0140067	24.69	0.000	.3183073	.3732125
Nikkei225						
L1.	-.0279048	.0081274	-3.43	0.001	-.0438342	-.0119754
_cons	.0004798	.0000999	4.80	0.000	.000284	.0006755
ARCH_FTSE100						
arch						
L1.	.0683949	.005021	13.62	0.000	.0585539	.0782358
garch						
L1.	.9213726	.0057566	160.06	0.000	.91009	.9326552
_cons	1.32e-06	1.94e-07	6.78	0.000	9.36e-07	1.70e-06
Nikkei225						
Dax30						
L1.	.0770417	.0179238	4.30	0.000	.0419117	.1121716
FTSE100						
L1.	.1498815	.0229469	6.53	0.000	.1049064	.1948567
SP500						
L1.	.3973318	.0178557	22.25	0.000	.3623353	.4323283
Nikkei225						
L1.	-.0981565	.0129139	-7.60	0.000	-.1234674	-.0728457
_cons	.0001054	.000143	0.74	0.461	-.0001749	.0003856
ARCH_Nikk~225						
arch						
L1.	.0942441	.007749	12.16	0.000	.0790563	.109432
garch						
L1.	.8899	.0085804	103.71	0.000	.8730828	.9067172
_cons	3.60e-06	5.85e-07	6.15	0.000	2.45e-06	4.74e-06

SP500						
Dax30						
L1.	.0251542	.0131107	1.92	0.055	-.0005423	.0508507
FTSE100						
L1.	.0079494	.0170475	0.47	0.641	-.0254631	.0413619
SP500						
L1.	-.0359037	.0150903	-2.38	0.017	-.0654801	-.0063273
Nikkei225						
L1.	-.0141287	.0079118	-1.79	0.074	-.0296355	.0013781
_cons	.0006734	.0001002	6.72	0.000	.000477	.0008698
ARCH_SP500						
arch						
L1.	.0555745	.0083194	6.68	0.000	.0392688	.0718802
garch						
L1.	1.206581	.1205421	10.01	0.000	.9703229	1.442839
L2.	-.2702983	.1124563	-2.40	0.016	-.4907086	-.0498879
_cons	1.02e-06	1.94e-07	5.28	0.000	6.43e-07	1.40e-06
corr(Dax30, FTSE100)	.8210761	.0199397	41.18	0.000	.781995	.8601571
corr(Dax30, Nikkei225)	.2301334	.0538521	4.27	0.000	.1245853	.3356816
corr(Dax30, SP500)	.5798921	.0373311	15.53	0.000	.5067245	.6530598
corr(FTSE100, Nikkei225)	.2223303	.0541032	4.11	0.000	.1162901	.3283706
corr(FTSE100, SP500)	.5539692	.0393262	14.09	0.000	.4768913	.6310471
corr(Nik~225, SP500)	.152188	.0552758	2.75	0.006	.0438494	.2605267
Adjustment						
lambda1	.0116965	.000875	13.37	0.000	.0099815	.0134115
lambda2	.9848049	.0011414	862.80	0.000	.9825678	.987042

```
. estimates store DCCRefined2

. lrtest DCC DCCRefined2
```

Likelihood-ratio test	LR chi2(1) =	5.12
(Assumption: DCC nested in DCCRefined2)	Prob > chi2 =	0.0236

```
. estimates stats DCC DCCRefined1 DCCRefined2
```

Akaike´s information criterion and Bayesian information criterion

Model	Obs	ll(null)	ll(model)	df	AIC	BIC
DCC	5,827	.	76682.15	40	-153284.3	-153017.5
DCCRefined1	5,827	.	76685.71	43	-153285.4	-152998.6
DCCRefined2	5,827	.	76684.71	41	-153287.4	-153013.9

Note: N=Obs used in calculating BIC; see [R] BIC note.

The model stops at a log-likelihood value of 76,684.71, but compared with the previous DCC model, here we have added just an extra parameter. Thus, the likelihood-ratio test indicates that the DCC model including an extra GARCH parameter for modeling the volatility of FTSE 100 contributes in a significant way to model the time structure of our data. The p-value associated with the likelihood-ratio test is now equal to 0.02, and therefore we can reject, with sufficiently strong evidence, the null hypothesis that the two models (restricted and unrestricted) are equal from a goodness-of-fit point of view. Thus, we choose the unrestricted model, the one with one extra parameter.

Dynamic conditional correlation Tse and Tsui (DCCT)

The second specification of DCC model is the one proposed by Tse and Tsui (2002). This model too relies on the same decomposition of the variance–covariance already illustrated when presenting the DCC model by Engle (2002).

$$H_t = D_t R_t D_t$$

The main difference between the Engle (2002) and Tse and Tsui (2002) models is the way the conditional correlation matrix R_t is modeled. In fact, Tse and Tsui (2002) make the assumption that elements of the matrix R_t follow an ARMA process:

$$R_t = (1 - \lambda_1 - \lambda_2) R + \lambda_1 \Psi_{t-1} + \lambda_2 R_{t-1} \qquad (4.17)$$

where λ_1 and λ_2 are nonnegative, time-invariant parameters satisfying the stationarity condition $\lambda_1 + \lambda_2 < 1$. Given that λ_1 and λ_2 are scalars, this implies that all the correlations follow the same dynamic, exactly as happened in the DCC model by Engle (2002), where the two parameters λ_1 and λ_2 govern the dynamic of correlations. The matrix $R(m \times m)$ in (4.17) is the long-run correlation matrix, which is guaranteed to be positive semidefinite. Finally, $\Psi(m \times m)$ is the correlation matrix of standardized residuals whose ijth element takes the form

$$\Psi_{ij,t-1} = \frac{\sum_{l=1}^{L} \dfrac{\varepsilon_{i,t-l}}{\sqrt{h_{ii,t-l}}} \dfrac{\varepsilon_{j,t-l}}{\sqrt{h_{jj,t-l}}}}{\sqrt{\sum_{l=1}^{L} \left(\dfrac{\varepsilon_{i,t-l}}{\sqrt{h_{ii,t-l}}} \right)^2 \sum_{l=1}^{L} \left(\dfrac{\varepsilon_{j,t-l}}{\sqrt{h_{jj,t-l}}} \right)^2}}$$

$L \geq m$ ensures that $\Psi_{ij,t}$ is positive semidefinite.

Regarding the formulation of the conditional correlation matrix, according to Engle (2002), conditional correlations follow a univariate GARCH model, while according to Tse and Tsui (2002), conditional correlations are a weighted sum of past correlations.

We now fit the DCCT model by using the same specification that we used for the DCCE model.

```
. mgarch vcc (Dax30 = L.Dax30 L.FTSE100 L.SP500 L.Nikkei225, arch(1) garch(1))
> (FTSE100 = L.Dax30 L.FTSE100 L.SP500 L.Nikkei225, arch(1) garch(1))
> (Nikkei225 = L.Dax30 L.FTSE100 L.SP500 L.Nikkei225, arch(1) garch(1))
> (SP500 = L.Dax30 L.FTSE100 L.SP500 L.Nikkei225, arch(1) garch(1/2)), nolog
Varying conditional correlation MGARCH model
```

Sample: 03jan1991 - 30dec2013

				Number of obs	=	5,827
Distribution: Gaussian				Wald chi2(16)	=	2175.49
Log likelihood = 76733.32				Prob > chi2	=	0.0000

	Coef.	Std. Err.	z	P>|z|	[95% Conf. Interval]	
Dax30						
Dax30						
L1.	-.1135001	.0176419	-6.43	0.000	-.1480775	-.0789227
FTSE100						
L1.	-.0423042	.0213724	-1.98	0.048	-.0841934	-.000415
SP500						
L1.	.3882836	.0171655	22.62	0.000	.3546398	.4219274
Nikkei225						
L1.	-.0155158	.0100691	-1.54	0.123	-.0352508	.0042192
_cons	.0008307	.0001258	6.60	0.000	.0005842	.0010772
ARCH_Dax30						
arch						
L1.	.0571081	.0041496	13.76	0.000	.048975	.0652412
garch						
L1.	.9280548	.0052652	176.26	0.000	.9177353	.9383744
_cons	2.14e-06	2.79e-07	7.65	0.000	1.59e-06	2.68e-06
FTSE100						
Dax30						
L1.	-.0668714	.0131189	-5.10	0.000	-.0925839	-.0411589
FTSE100						
L1.	-.092314	.0174654	-5.29	0.000	-.1265456	-.0580824
SP500						
L1.	.3477459	.0138447	25.12	0.000	.3206107	.374881
Nikkei225						
L1.	-.0292403	.0080468	-3.63	0.000	-.0450117	-.0134689
_cons	.0004927	.0000992	4.97	0.000	.0002983	.0006871
ARCH_FTSE100						
arch						
L1.	.0598549	.0044687	13.39	0.000	.0510964	.0686135
garch						
L1.	.9279083	.0053615	173.07	0.000	.9174	.9384166
_cons	1.11e-06	1.69e-07	6.58	0.000	7.80e-07	1.44e-06

Nikkei225						
Dax30						
L1.	.0746545	.01789	4.17	0.000	.0395907	.1097182
FTSE100						
L1.	.1483266	.0228553	6.49	0.000	.103531	.1931222
SP500						
L1.	.4028095	.0178941	22.51	0.000	.3677377	.4378814
Nikkei225						
L1.	-.0961706	.0129116	-7.45	0.000	-.1214768	-.0708644
_cons	.0001088	.0001431	0.76	0.447	-.0001717	.0003894
ARCH_Nikk~225						
arch						
L1.	.0912404	.007503	12.16	0.000	.0765349	.105946
garch						
L1.	.8924415	.0083897	106.37	0.000	.8759979	.908885
_cons	3.44e-06	5.64e-07	6.09	0.000	2.33e-06	4.54e-06
SP500						
Dax30						
L1.	.0233116	.013001	1.79	0.073	-.0021699	.0487932
FTSE100						
L1.	.0101257	.016904	0.60	0.549	-.0230054	.0432569
SP500						
L1.	-.0361813	.0150019	-2.41	0.016	-.0655845	-.0067782
Nikkei225						
L1.	-.0152176	.0078731	-1.93	0.053	-.0306486	.0002135
_cons	.000672	.0001003	6.70	0.000	.0004754	.0008685
ARCH_SP500						
arch						
L1.	.051923	.0079945	6.49	0.000	.0362541	.067592
garch						
L1.	1.222177	.1225847	9.97	0.000	.9819156	1.462439
L2.	-.284033	.1144855	-2.48	0.013	-.5084206	-.0596455
_cons	9.84e-07	1.89e-07	5.22	0.000	6.14e-07	1.35e-06
corr(Dax30, FTSE100)	.8807403	.0242385	36.34	0.000	.8332338	.9282468
corr(Dax30, Nikkei225)	.3162826	.0595738	5.31	0.000	.1995201	.433045
corr(Dax30, SP500)	.734496	.0474448	15.48	0.000	.6415059	.8274862
corr(FTSE100, Nikkei225)	.2939437	.0589945	4.98	0.000	.1783167	.4095708

corr(FTSE100, SP500)	.7123827	.0514333	13.85	0.000	.6115753	.8131902
corr(Nik~225, SP500)	.2489124	.0620554	4.01	0.000	.127286	.3705387
Adjustment						
lambda1	.0054936	.0007229	7.60	0.000	.0040767	.0069104
lambda2	.9930198	.000939	1057.49	0.000	.9911793	.9948603

```
. estimates store DCCT_Gaussian
```

The structure of the output is exactly the same as the one obtained in the DCCE model reported in the previous section. The log likelihood of the DCCT model is equal to 76,733.32, while the DCCE model stopped at a log likelihood of 76,684.71. Therefore, given that the two models have the same number of parameters, the best model is the one by Tse and Tsui, without the need to conduct a likelihood-ratio test. We report the variances (see figure 4.9) and correlations (see figure 6.1) of the Tse and Tsui model.

```
. predict H_DCCT*, variance
. tsset date
        time variable:  date, 1/2/1991 to 12/30/2013, but with gaps
               delta:  1 day
. tsline H_DCCT_Dax30_Dax30, name(VolDCCTDax30)
. tsline H_DCCT_FTSE100_FTSE100, name(VolDCCTFTSE100)
. tsline H_DCCT_Nikkei225_Nikkei225, name(VolDCCTNikkei225)
. tsline H_DCCT_SP500_SP500, name(VolDCCTSP500)
. graph combine VolDCCTDax30 VolDCCTFTSE100 VolDCCTNikkei225 VolDCCTSP500
```

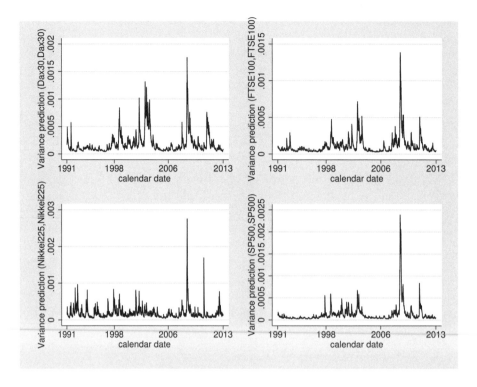

Figure 4.9. Variances from the DCCT model

```
. predict corr_DCCT*, correlation
. tsline corr_DCCT_FTSE100_Dax30, ylabel(0(.2)1) name(corrdcctftsedax)
. tsline corr_DCCT_Nikkei225_Dax30, ylabel(0(.2)1) name(corrdcctnikkeidax)
. tsline corr_DCCT_SP500_Dax30, ylabel(0(.2)1) name(corrdcctspdax)
. tsline corr_DCCT_Nikkei225_FTSE100, ylabel(0(.2)1) name(corrdcctnikkeiftse)
. tsline corr_DCCT_SP500_FTSE100, ylabel(0(.2)1) name(corrdcctspftse)
. tsline corr_DCCT_SP500_Nikkei225, ylabel(0(.2)1) name(corrdcctspnikkei)
. graph combine corrdcctftsedax corrdcctnikkeidax corrdcctspdax corrdcctnikkeiftse
> corrdcctspftse corrdcctspnikkei
```

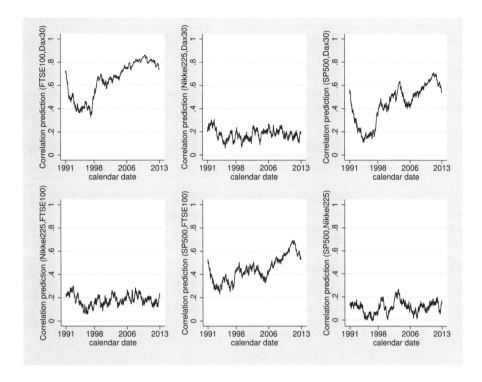

Figure 4.10. Correlations from the DCCT model

We have described most of the features that Stata offers to fit MGARCH models. The final issue to discuss is the choice of the multivariate distribution under which to carry out the estimation. We can choose between the multivariate Gaussian and the multivariate Student's t distribution.

We proceed by fitting the model reported above under the assumption that innovations follow a multivariate Student's t distribution. Because we have already obtained the estimates under the assumption of a multivariate Gaussian distribution, we use these estimates as starting values for the estimation under the multivariate Student's t distribution. This will speed up the estimation procedure because we can suppose that parameters obtained under the Gaussianity assumption are a good starting point for the optimization problem. If we did not specify any starting value, the estimation process would spend some time looking for its own starting points.

To specify our own set of starting points, we create a vector containing parameter estimates for the DCC under the Gaussianity assumption that we can recall using the syntax e(b). We need to add an extra parameter for the degrees of freedom of the multivariate Student's t distribution, and we set this equal to 30.

```
. tsset newdate
        time variable:  newdate, 02jan1991 to 30dec2013
                delta:  1 day
. matrix parameters=[e(b),30]
```

We create a new vector named **parameters** using the keyword **matrix**, concatenating the parameters stored in the vector **e(b)** and a scalar, 30.

Once we have the vector of starting values, we fit the DCC model under multivariate Student's t, initializing the estimation procedure with those values by using the keyword **from**.

```
. mgarch vcc (Dax30 = L.Dax30 L.FTSE100 L.SP500 L.Nikkei225, arch(1) garch(1))
> (FTSE100   = L.Dax30 L.FTSE100 L.SP500 L.Nikkei225, arch(1) garch(1))
> (Nikkei225 = L.Dax30 L.FTSE100 L.SP500 L.Nikkei225, arch(1) garch(1))
> (SP500 = L.Dax30 L.FTSE100  L.SP500 L.Nikkei225, arch(1) garch(1/2)),
> distribution(t) from(parameters) nolog
Varying conditional correlation MGARCH model
```

```
Sample: 03jan1991 - 30dec2013                Number of obs   =       5,827
Distribution: t                              Wald chi2(16)   =     2230.62
Log likelihood =  77255.53                   Prob > chi2     =      0.0000
```

	Coef.	Std. Err.	z	P>|z|	[95% Conf. Interval]	
Dax30						
Dax30						
L1.	-.1201215	.0165316	-7.27	0.000	-.1525228	-.0877201
FTSE100						
L1.	-.0361937	.0200756	-1.80	0.071	-.0755412	.0031538
SP500						
L1.	.3754259	.0164229	22.86	0.000	.3432376	.4076142
Nikkei225						
L1.	-.0259908	.0093627	-2.78	0.006	-.0443413	-.0076403
_cons	.0008305	.0001174	7.08	0.000	.0006005	.0010605
ARCH_Dax30						
arch						
L1.	.0481563	.0040639	11.85	0.000	.0401912	.0561213
garch						
L1.	.9420708	.0047184	199.66	0.000	.9328229	.9513188
_cons	1.22e-06	2.08e-07	5.88	0.000	8.15e-07	1.63e-06
FTSE100						
Dax30						
L1.	-.0805236	.012719	-6.33	0.000	-.1054524	-.0555947
FTSE100						
L1.	-.0760358	.0170293	-4.46	0.000	-.1094127	-.042659

SP500						
L1.	.3323518	.0135295	24.57	0.000	.3058345	.358869
Nikkei225						
L1.	-.0337213	.0077718	-4.34	0.000	-.0489537	-.0184888
_cons	.0005106	.000096	5.32	0.000	.0003225	.0006987
ARCH_FTSE100						
arch						
L1.	.0469445	.0041994	11.18	0.000	.0387139	.055175
garch						
L1.	.9433863	.0049508	190.55	0.000	.9336829	.9530897
_cons	8.63e-07	1.49e-07	5.79	0.000	5.71e-07	1.16e-06
Nikkei225						
Dax30						
L1.	.0756805	.0173679	4.36	0.000	.0416401	.1097209
FTSE100						
L1.	.1487646	.0224668	6.62	0.000	.1047304	.1927988
SP500						
L1.	.3991064	.0177417	22.50	0.000	.3643333	.4338795
Nikkei225						
L1.	-.1134056	.0122428	-9.26	0.000	-.1374011	-.0894101
_cons	.0000273	.0001366	0.20	0.842	-.0002404	.000295
ARCH_Nikk~225						
arch						
L1.	.0692826	.0065283	10.61	0.000	.0564874	.0820778
garch						
L1.	.9209628	.0071306	129.16	0.000	.9069872	.9349385
_cons	2.01e-06	4.35e-07	4.63	0.000	1.16e-06	2.87e-06
SP500						
Dax30						
L1.	.0152844	.0122787	1.24	0.213	-.0087815	.0393502
FTSE100						
L1.	.0154102	.0160667	0.96	0.337	-.01608	.0469005
SP500						
L1.	-.0363608	.0143789	-2.53	0.011	-.0645428	-.0081788
Nikkei225						
L1.	-.0155239	.0074763	-2.08	0.038	-.0301772	-.0008706
_cons	.000709	.0000952	7.45	0.000	.0005225	.0008955

```
ARCH_SP500
        arch
         L1.     .0390478    .0079403     4.92   0.000     .0234851    .0546106

       garch
         L1.    1.331517      .13972      9.53   0.000    1.057671    1.605363
         L2.    -.3778037   .1314131     -2.87   0.004    -.6353686   -.1202387

       _cons    6.49e-07    1.57e-07     4.12   0.000     3.40e-07    9.57e-07

  corr(Dax30,
     FTSE100)   .8974011    .0284186    31.58   0.000     .8417018    .9531005
  corr(Dax30,
    Nikkei225)  .245601      .06507      3.77   0.000     .1180661    .373136
  corr(Dax30,
       SP500)   .7398684    .0544607    13.59   0.000     .6331273    .8466095
corr(FTSE100,
    Nikkei225)  .2422963    .0651667     3.72   0.000     .114572     .3700207
corr(FTSE100,
       SP500)   .7092343    .0583177    12.16   0.000     .5949337    .8235349
corr(Nik~225,
       SP500)   .2409208    .068883      3.50   0.000     .1059125    .375929

Adjustment
     lambda1    .0055255    .0007318     7.55   0.000     .0040912    .0069597
     lambda2    .9930641    .0009421  1054.10   0.000     .9912176    .9949106

df
       _cons    9.207344     .453134    20.32   0.000     8.319218    10.09547
```

. estimates store DCCT_StudentT

The degrees of freedom is equal to 9.21, suggesting that the fitted multivariate Student's t distribution is quite far from the multivariate normal and is characterized by moderately fat tails. We note that the other parameters are not too far from those obtained by estimating the DCC model under the multivariate normal distribution. The optimization procedure stops at a value equal to 77,255.53 while when working with the Gaussian distribution the log likelihood was equal to 76,733.32. Although the Student's t distribution is quite evidently a better fit over the Gaussian one, we want to formally compare the two DCC models by applying the likelihood-ratio test.

. lrtest DCCT_StudentT DCCT_Gaussian

```
Likelihood-ratio test                              LR chi2(1)  =    1044.42
(Assumption: DCCT_Gaussian nested in DCCT_StudentT)  Prob > chi2 =     0.0000
```

. estimates stats DCCT_Gaussian DCCT_StudentT

Akaike's information criterion and Bayesian information criterion

Model	Obs	ll(null)	ll(model)	df	AIC	BIC
DCCT_Gauss~n	5,827	.	76733.32	41	-153384.6	-153111.2
DCCT_Stude~T	5,827	.	77255.53	42	-154427.1	-154146.9

Note: N=Obs used in calculating BIC; see [R] BIC note.

As expected, the likelihood-ratio test strongly rejects the equivalence of the two models in favor of the unrestricted one, the DCC under multivariate Student's t distribution. Coherently, the information criteria are lower for the model estimated under multivariate Student's t distribution compared with the multivariate Gaussian.

Prediction

A final feature we discuss about MGARCH models is the possibility to obtain some forecasts. We use the `predict` command to do this; in this framework, we obtain forecasts not only for the mean and the conditional variance but also for the conditional covariances and correlations.

We consider the DCCT specification presented above under the Student's t assumption. We obtain dynamic forecasts for both conditional variances and correlations for the next 100 time periods into the future. With this purpose, we start by adding 100 new rows to our dataset by using the syntax `tsappend, add(100)`, where `add()` takes as an argument the number of periods for which we want to obtain forecasts. The next step requires the `predict` command, where we specify the `dynamic(mdy(12,30,2013))` option taking as an argument the time point from which we want to obtain forecasts.

```
. tsset date
        time variable:   date, 1/2/1991 to 12/30/2013, but with gaps
                delta:   1 day

. tsappend, add(100)

. predict H_Forecast*, variance dynamic(mdy(12,30,2013))

. tsline H_Forecast_SP500_SP500 H_Forecast_FTSE100_FTSE100
> H_Forecast_Dax30_Dax30 H_Forecast_Nikkei225_Nikkei225
> if date>=mdy(12,31,2011),
> tlabel(01jan2012 01jan2013 01jan2014 01jan2015, format(%tdnn/dd/ccYY))
> tline(30dec2013) legend(rows(4))
```

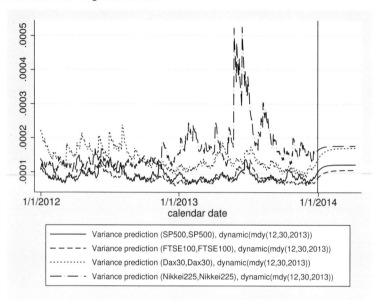

Figure 4.11. Variances forecast from the DCCT model

In figure 4.11, we highlighted the end of the historical data from the forecasts by drawing a vertical line with the `tline()` option. As we have stated before, the forecasts that we can obtain from a GARCH model are meaningful just for the short horizon.

We proceed in a similar way to obtain forecasts for correlations (see figure 6.1).

```
. predict corr_For*, correlation dynamic(mdy(12,30,2013))
(no lags of dependent variables on rhs; option dynamic() ignored)

. tsline corr_For_SP500_Dax30 corr_For_SP500_FTSE100 corr_For_SP500_Nikkei225
> if date >= mdy(12,31,2011),
> tlabel(01jan2012 01jan2013 01jan2014 01jan2015, format(%tdnn/dd/ccYY))
> tline(30dec2013) legend(rows(3))
```

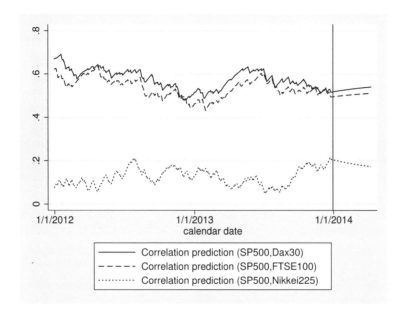

Figure 4.12. Correlations forecast from the DCCT model

As with the variances, when we consider correlations, we can see as we move away from the end of the historical series that forecasts tend to a long-run value.

In addition, we may require forecasts to be computed just for a specific equation by indicating the name of the equation in which we are interested using the `equation(`*y1*`)` option.

4.5   Final remarks

In this chapter, we presented and illustrated the main features of MGARCH models belonging to the category of nonlinear combinations of univariate GARCH. We highlighted the importance of allowing correlations to be time-varying and therefore to move from the CCC model by Bollerslev (1990) to some more flexible models. Both DCC models, by Engle (2002) and by Tse and Tsui (2002), show some good properties: they both recognize the dynamic nature of correlations while remaining feasible from a computational point of view, even when a quite large number of time series is considered. The two-step estimation procedure ensures the feasibility of the optimization problem.

However, the DCC models present some noticeable drawbacks that recent literature has addressed. The first drawback is about the assumption that all the elements of the covariance matrix follow the same dynamic, because their pattern is defined by two scalar values—λ_1 and λ_2 in (4.13) for the DCCE, and λ_1 and λ_2 in (4.17) for the DCCT. Of course, this restriction can be quite unrealistic, especially when we are working with a large number of assets.

When the number of assets considered increases, it becomes more important to be able to handle the problem from a mathematical point of view by imposing some constraints. A model that relaxes this assumption was proposed by Cappiello, Engle, and Sheppard (2006), the asymmetric generalized dynamic conditional correlation (AGDCC) model. In this model, instead of two coefficients λ_1 and λ_2, the dynamic of matrix Q depends on two matrices, A and B, each of them collecting specific coefficients for all the elements of Q, allowing a much richer flexibility. As usual, higher flexibility comes at the cost of longer computational time.

Another interesting extension of the DCC model deals with the assumption that the long-run or unconditional covariance matrix is constant throughout time and regardless of the time span analyzed. Once again, this assumption is hardly supported by the data, especially when the time period considered gets quite long. Colacito, Engle, and Ghysels (2011) deal explicitly with this limitation of the DCC model and introduce the DCC-MIDAS model with mixed data sampling (MIDAS), a general technique introduced by Ghysels, Santa-Clara, and Valkanov (2004, 2005, 2006) and Ghysels, Sinko, and Valkanov (2007). With this model, we can mix data recorded at two different time frequencies, for instance, at daily and monthly frequency. This feature allows us to come up with a unique model that is able to describe two dynamics of correlations, one pertaining to the high frequency domain and the other to the low frequency domain. Moreover, the two time scales can be associated to alternative explanatory variables, for instance, macroeconomic and market momentum variables. For further details, see the references aforementioned.

5 Risk management

5.1 Introduction

Risk management is the intersection of several quantitative disciplines dealing with the theoretical and practical evaluation of risk. Many alternative definitions of risk occur in finance, with the most relevant being market risk, credit risk, operational risk, liquidity risk, and model risk.

We can define market risk as the risk of a decrease in the value of a financial position due to unforeseen changes in the value of the underlying components on which that position depends, such as stock and bond prices or exchange rates. In this chapter, we focus mainly on this kind of risk.

However, for completeness in what follows, we briefly define the other categories of risk. Credit risk arises on the uncertainty that the counterpart will meet its financial obligations. For instance, if the counterpart is a bond issuer, the risk is that he will default on the repayment of coupons or principal. In addition to that, if the issued bonds are rated, we can be exposed even to the risk of downgrading, which is the risk that a rating agency will decrease its overall judgment of debt-issuer creditworthiness, directly affecting the market price of the bond.

The Basel Committee on Banking Supervision in 2004 introduced a new risk category, operational risk, defined as the risk of direct or indirect losses resulting from inadequate or failed internal processes, people, and systems or from external events. This definition includes legal risk, but excludes strategic and reputational risk. Examples of losses falling within this category are fraud, losses due to IT and power failures, errors in settlements of transactions, litigation, and losses due to external events like flooding, fire, earthquake, or terrorism.

Model risk refers to the risk occurring when using a misspecified financial model for undertaking some decisions, for instance, measuring risk or determining the price of financial securities. The first document formally defining model risk is the Supervisory Guidance on Model Risk Management published by the Office of the Comptroller of the Currency and the Federal Reserve in 2011–2012, providing guidelines for identifying and managing this kind of risk.

Finally, liquidity risk refers to the lack of marketability of some financial instruments, that is, the impossibility to trade it in a short time. Liquidity risk is especially important during periods of financial turmoil, because of rising concerns about the evolution of the market prices of some assets. An example is the burst of the 2010 European sovereign crisis, when Greek bonds experienced a dramatic widening of their bid-ask spread[1] as a sign of the asymmetry in the number of financial investors willing to buy and sell these bonds. During that period, owners of Greek bonds found it difficult to sell these bonds because of few buyers for these assets on the financial markets. These bond owners thus experienced a liquidity risk.

Given that the focus of this chapter is on measuring market risk, we need to introduce the most common measure adopted by financial institutions to evaluate the market risk exposure of their trading portfolios, the known value-at-risk (VaR). We define VaR as the maximum loss that a portfolio will experience over a given time horizon with $\alpha\%$ probability. Therefore, from a mathematical point of view, VaR is the $(1-\alpha)$th quantile of the profit and loss (P&L) distribution over a target horizon.

Over the last decade, a wide array of parametric methods (for instance, risk metrics and GARCH) and nonparametric methods (for instance, historical and Monte Carlo simulations) have been proposed to quantify VaR. Because financial institutions are required to hold regulatory capital based on their VaR forecasts, much effort has been devoted to developing techniques aimed at validating their measure of market risk. Hence, if on the one hand it is relevant for banks to implement accurate VaR models, then on the other hand they need to use sound statistical backtest procedures to validate them.

To address these issues, we discuss in this chapter the alternative methodologies to correctly quantify VaR and the statistical tests expressly designed to assess the appropriateness of a VaR measure.

5.2   Loss

The loss of a portfolio over the period $[t, t + \Delta]$ is defined as follows:

$$L_{[t,t+\Delta]} = -\left\{V\left(t, t + \Delta\right) - V\left(t\right)\right\}$$

where $V(t)$ is the value of a portfolio at time t, and Δ is the time horizon that is relevant to compute the losses (when considering daily data, we will set $\Delta = 1/365$, while when considering weekly data, we will set $\Delta = 1/52$).

Having defined the portfolio loss, the first issue that risk managers have to face is how to map the value of their portfolio to d risk factors $Z_t = (Z_{t,1}, \ldots, Z_{t,d})$ in such a way that, by obtaining a future distribution for these risk factors, we will be able to directly obtain a future distribution for the value of the portfolio. In mathematical terms, addressing this point means to define a form for function f:

$$V_t = f\left(t, Z_t\right) \tag{5.1}$$

The choice of risk factors is a tough job because it largely depends on the dimensionality of the portfolio we are analyzing. For a small portfolio, we can use asset returns so that

---

1. Bid-ask spread is the difference in price between the highest price that a buyer is willing to pay for an asset and the lowest price for which a seller is willing to sell it.

the value of each asset at time t is obtained using asset price formulas; for example, the value of a stock at time $t+1$ can be computed as $S_{t+1} = S_t e^{r_{t,t+1}}$. When the portfolio gets quite big, it can be meaningful to carry out a preliminary principal component analysis to identify the main risk factors affecting the portfolio value and then map the assets to these risk factors. Following this approach, we are able to deal with fewer risk factors and therefore overcome the portfolio dimensionality.

Having in hand the link between risk factors and portfolio value, we can be more specific about the loss distribution. We are interested in the conditional loss distribution, which is defined as the loss distribution given all available information up to and including time s. In mathematical terms, let us consider the matrix of risk-factor changes X_t defined as $X_t = Z_t - Z_{t-1}$ and assumed to follow a stationary distribution F_x on R^d, and let us consider the σ-algebra generated by the historical time series of risk-factor changes, $\mathcal{F}_t = \sigma\{(X_s : s \le t)\}$. We define the conditional loss distribution L as

$$F_{L_{t+1}|\mathcal{F}_t(l)} = P\{l_t(X_{t+1}) \le l|\mathcal{F}_t\} = P(L_{t+1} \le l|\mathcal{F}_t)$$

where l is some loss threshold.

5.3   Risk measures

Definition 5.1. Let M be the set of random variables representing portfolio losses L over a period of time and let ρ be a risk measure. We say that a risk measure is coherent when it satisfies the following properties:

1. Translation invariance. For all $L \in M$ and $l \in \mathbb{R}$, it must follow that $\rho(L + l) = \rho(L) + l$. The intuitive understanding of this condition is that by adding or subtracting the amount l to our portfolio, we increase or reduce the overall risk L by the amount l.

2. Subadditivity. $\forall L_1, L_2 \in M$, we should verify that $\rho(L_1 + L_2) \le \rho(L_1) + \rho(L_2)$. This condition requires that the risk of a linear combination of two positions not be more risky than the sum of the risks associated with the two individual positions. This condition should hold true because of the risk diversification.

3. Positive homogeneity. $\forall \lambda \ge 0$, it must follow that $\rho(\lambda L) = \lambda \rho(L)$. The condition of positive homogeneity requires that the risk scale with the size of a position: when the size of a position makes it more difficult to sell the position in the market, that means it gets illiquid, and then the market risk associated with that position should consider even the liquidity risk.

4. Monotonicity. $\forall L_1, L_2 \in M$ s.t. $L_1 \le L_2$, we should verify that $\rho(L_1) \le \rho(L_2)$. The condition reads as follows: if the future net loss L_1 is smaller than L_2, then L_1 is less risky than L_2.

Although there are several risk measures, just a few of them belong to the class of coherent risk measures. For example, the standard deviation is not a coherent measure, even though it is one of the bedrocks of Markowitz theory, because it satisfies neither the translation property nor the monotonicity one. This measure attributes the same weights to positive and negative deviations from the mean; thus, it is unable to describe rare events. Therefore, standard deviation is a good risk measure just when the P&L distribution is symmetric around the mean; in the presence of highly skewed distributions, it could understate the risk.[2]

Another risk measure is the semivariance. This measure considers just the observations falling below a certain threshold θ, usually set equal to 0:

$$\text{semiSD} = \sum_{t=1}^{T} \frac{\{r_t I \left(r_t < \theta\right)\}^2}{T-1}$$

where $I(\cdot)$ is the indicator function, taking a value of 1 when the condition $r_t < \theta$ is met and a value of 0 otherwise.

5.4   VaR

The most famous measure of risk is the VaR, the value at risk. This measure was introduced in the early 1990s by J. P. Morgan, and its great popularity is attributable to the easiness of the risk concept it covers. The VaR measure answers the following question:

> What is the maximum loss my portfolio can experience in the next Δ days with a probability of α?

We can use VaR to quantify not only market risk but also other forms of risk a financial asset is exposed to, such as credit, operational, and liquidity risk. In addition to that, regulators frequently use estimates of VaR to determine minimum capital adequacy requirements.

We now give a formal definition of VaR.

Definition 5.2. Let $\alpha \in (0,1)$ be the confidence level at which we want to determine VaR. The α-level VaR is

$$\text{VaR}_\alpha = \inf \left\{ l \in \mathbb{R} : P\left(L < l\right) \leq \left(1 - \alpha\right) \right\} = \inf \left\{ l \in \mathbb{R} : F_L\left(l\right) \geq \alpha \right\} = F_L^{-1}(\alpha) \quad (5.2)$$

In (5.2), we define VaR as the lth quantile of the loss distribution, such that the $\text{VaR}_\alpha(L)$ over the horizon time Δ would be exceeded on average $100(1 - \alpha)$ times every 100Δ time periods. The confidence level α is typically a large number between 0.95 and 1, and the forecast horizon is usually set between 1 and 10 days.

The main drawback of VaR is that it does not provide any information about the severity of the loss once it occurs. VaR indicates the maximum loss that we could

---

2. A more technical drawback of standard deviation is that the second moment of the loss distribution is assumed to exist, which is not always the case (for example, the Cauchy distribution).

experience, with a specified confidence level and over a predefined time horizon, but it does not tell anything about the severity of the loss once the threshold is passed. To illustrate a problem with the VaR measure, consider two loss distributions with the same 1% quantile that have very different left-hand tails; F_{L1} has a long, thin left tail, and F_{L2} has a short, thick left tail. The VaR measures are the same even though some large losses occasionally occur under F_{L1} that have zero probability under F_{L2}. By limiting ourselves to VaR, we will not be able to identify which is the most risky asset.

An additional drawback of VaR, certainly not less relevant, is that it does not fulfill the subadditivity property, making VaR a noncoherent measure. To show this, let us consider F_{L1} and F_{L2} to be two loss distributions for two assets and F_L to be the loss distribution of the aggregated portfolio $L = L1 + L2$. The subadditivity condition requires that $q_\alpha(F_L) \leq q_\alpha(F_{L1}) + q_\alpha(F_{L2})$, with the q_αth quantile of the distribution. This condition is not guaranteed by the VaR, and therefore the risk of the aggregated portfolio is not always smaller or equal to the sum of the risk associated with the two single positions. In this sense, VaR neglects the diversification benefit that forms the basis of modern portfolio theory. Because it violates subadditivity, we can state that VaR is not a coherent risk measure.

5.4.1 VaR estimation

We can estimate VaR according to three alternative approaches:

1. parametric model
2. historical simulation
3. Monte Carlo simulation

Another common technique that we can adopt to estimate VaR relies on the extreme value theory. In fact, because VaR calculations are only concerned with the tails of a probability distribution, techniques from extreme value theory (EVT) may be particularly effective.

5.4.2 Parametric approach

A parametric VaR model assumes that returns follow a specific distribution, usually the normal, whose parameters are either estimated on the basis of historical data or set equal to some forward-looking measures, for instance, the implied volatility of options. The parametric approach is also known as the variance–covariance or delta-normal approach.

To formally introduce this method, let us assume that $X_{(t+1)}$ is a d-dimensional distribution of i.i.d. changes in risk factors following a multivariate normal distribution with mean vector $\mu(d \times 1)$ and variance–covariance matrix $\Sigma(d \times d)$. $X_{(t+1)} \sim N_d(\mu, \Sigma)$ drives the aggregated portfolio position L.

$$L_{t+1}^\Delta = l_t^\Delta (X_{t+1})$$

with $l_t(\cdot)$ being the function mapping risk factors X on the loss distribution L. The choice of the form of the function $l_t(\cdot)$ depends upon the assets composing the portfolio: if the portfolio includes only equity positions, then we can assume $l_t(\cdot)$ to be linear; when it also involves assets with nonlinear pay-off (for example, options), then we need $l_t(\cdot)$ to be nonlinear to correctly model the linkage between risk factors and portfolio loss. Assuming that we can describe the joint distribution of portfolio returns as a multivariate normal distribution with mean vector μ and variance–covariance matrix Σ, the loss function of a portfolio of d assets, all of them characterized by a linear pay-off and with vector weights w, takes the form

$$L_{t+1}^{\Delta} = l_t^{\Delta}\left(X_{t+1}\right) \sim N\left(w^{'}\mu, w^{'}\Sigma w\right) \tag{5.3}$$

From the loss distribution in (5.3), we can compute the VaR at the α confidence level as

$$
\begin{aligned}
\mathrm{VaR}_\alpha(L) &= \inf\left\{l | F_L(l) \geq \alpha\right\} \\
\mathrm{VaR}_\alpha(L) &= F_L^{-1}(\alpha) \\
\mathrm{VaR}_\alpha(L) &= w^{'}\mu + \left(w^{'}\Sigma w\right)\phi^{-1}(\alpha)
\end{aligned}
$$

where $\phi^{-1}(\alpha)$ is the αth quantile of the standard normal distribution.

More generally, if the loss distribution L follows a generic distribution F_L belonging to the location-scale family with location parameter μ and scale parameter σ^2, then we can compute the VaR as

$$\mathrm{VaR}_\alpha(L) = \mu + \sigma q_\alpha \tag{5.4}$$

where q_α is the αth quantile of the standardized distribution of L.

We now consider the daily returns of the S&P 500 over the years 1991–2013 and compute the VaR under the normality assumption. To do that, we first have to compute the mean and the standard deviation for the S&P 500 returns by using the generalized method of moments (GMM) method. As a quick reminder, method of moments estimators are found by equating the first k sample moments to the corresponding k population moments and solving the resulting system of simultaneous equations. We adopt the GMM method because it allows us also to obtain an estimate of the standard errors for both mean and variance, and therefore we will be able to compute standard errors and confidence intervals even for the VaR estimates.

```
. use http://www.stata-press.com/data/feus/index
. quietly summarize SP500
. local n = r(N)
. scalar mean = r(mean)
```

```
. gmm (SP500 - {mu})((SP500 - {mu})^2 - (`n´-1)/(`n´)*{v}), onestep
> winitial(identity) vce(robust)
Step 1
Iteration 0:   GMM criterion Q(b) =   1.010e-07
Iteration 1:   GMM criterion Q(b) =   6.925e-15
Iteration 2:   GMM criterion Q(b) =   5.675e-39

note: model is exactly identified

GMM estimation

Number of parameters =    2
Number of moments    =    2
Initial weight matrix: Identity              Number of obs   =      5,828
```

	Coef.	Robust Std. Err.	z	P>|z|	[95% Conf. Interval]	
/mu	.0002885	.0001512	1.91	0.056	-7.89e-06	.0005848
/v	.0001333	5.80e-06	22.98	0.000	.0001219	.0001446

```
Instruments for equation 1: _cons
Instruments for equation 2: _cons
```

We specify `quietly` just before `summarize` to suppress the output. Then we use the `gmm` command to obtain estimates for mean `mu` and variance `v`. The `gmm` command is followed by two equations in brackets reporting suitable expressions for the two moments we want to estimate. Indicating with μ the population mean and with y a random sample from the population, the equation for the mean is

$$E(y) - \mu = 0$$

$$\frac{1}{T} \sum_{t=1}^{t} (y_t - \widehat{\mu}) = 0$$

where $\widehat{\mu}$ is the GMM estimator of the true population parameter μ.

The equation for the variance is

$$E\left\{(y - \mu)^2\right\} - \sigma^2 = 0$$

$$\frac{1}{T-1} \sum_{t}^{T} \left\{(y_t - \widehat{\mu})^2 - \sigma^2\right\}$$

We use three options with `gmm`. `onestep` indicates that we want to use a one-step GMM estimator. The main difference with the two-step GMM estimator, the default option, is that here once parameters have been estimated based on an initial weight matrix, no further updating of the weight matrix is performed [except when calculating the appropriate variance–covariance estimator (VCE) matrix]. We choose to use the identity matrix, `winitial(identity)`, as the weight matrix to obtain the first-step parameter estimates. And finally, we compute robust standard errors by specifying `vce(robust)`.

Now we can compute the VaR for three confidence levels, 95%, 97%, and 99%, applying (5.4). With this purpose, we are going to use the `nlcom` command because we need to estimate a nonlinear combination of the estimated parameters. `nlcom` is a command specifically designed to compute point estimates, standard errors, test statistics, significance levels, and confidence intervals for nonlinear combinations of parameter estimates after any Stata estimation command.

```
. nlcom (VaR95: _b[mu:_cons] + sqrt(_b[v:_cons])*invnormal(0.05))
> (VaR97: _b[mu:_cons] + sqrt(_b[v:_cons])*invnormal(0.03))
> (VaR99: _b[mu:_cons] + sqrt(_b[v:_cons])*invnormal(0.01)), noheader
```

	Coef.	Std. Err.	z	P>|z|	[95% Conf.	Interval]
VaR95	-.0187003	.0004502	-41.54	0.000	-.0195826	-.017818
VaR97	-.0214241	.0005064	-42.30	0.000	-.0224166	-.0204315
VaR99	-.0265677	.0006142	-43.26	0.000	-.0277715	-.0253639

The VaR at 1 day for a position on the S&P 500 is -1.87% at the 95% confidence level, with a confidence interval of $[-1.96\%, -1.78\%]$. At the 97% confidence level, it is -2.14% $[-2.24\%, -2.04\%]$, and at the 99% level, is it -2.66% $[-2.78\%, -2.54\%]$. Note that daily returns are in percentages.

If we are interested in obtaining the VaR at 1 week, we just have to multiply the daily VaR by $\sqrt{5}$, the number of working days in a week, while to obtain the VaR at 1 month, we would multiply the daily VaR by $\sqrt{22}$, the average number of working days in a month.

```
. * weekly VaR
. nlcom (VaR95: (_b[mu:_cons] + sqrt(_b[v:_cons])*invnormal(0.05))*sqrt(5))
> (VaR97: (_b[mu:_cons] + sqrt(_b[v:_cons])*invnormal(0.03))*sqrt(5))
> (VaR99: (_b[mu:_cons] + sqrt(_b[v:_cons])*invnormal(0.01))*sqrt(5)), noheader
```

	Coef.	Std. Err.	z	P>|z|	[95% Conf.	Interval]
VaR95	-.0418151	.0010066	-41.54	0.000	-.0437881	-.0398422
VaR97	-.0479057	.0011324	-42.30	0.000	-.0501252	-.0456862
VaR99	-.0594072	.0013734	-43.26	0.000	-.062099	-.0567154

```
. * monthly VaR
. nlcom (VaR95: (_b[mu:_cons] + sqrt(_b[v:_cons])*invnormal(0.05))*sqrt(22))
> (VaR97: (_b[mu:_cons] + sqrt(_b[v:_cons])*invnormal(0.03))*sqrt(22))
> (VaR99: (_b[mu:_cons] + sqrt(_b[v:_cons])*invnormal(0.01))*sqrt(22)), noheader
```

	Coef.	Std. Err.	z	P>|z|	[95% Conf.	Interval]
VaR95	-.0877122	.0021116	-41.54	0.000	-.0918508	-.0835736
VaR97	-.1004878	.0023753	-42.30	0.000	-.1051434	-.0958322
VaR99	-.1246136	.0028808	-43.26	0.000	-.1302599	-.1189673

As discussed in chapter 1, financial returns often fail to meet the normality assumption. For this reason, we now consider a fatter-tail distribution, namely, the Student's t with 5 degrees of freedom. We compute VaR for the three confidence levels 95%, 97%, and 99%:

```
. * parametric VaR on SP 500 Student t distribution
. nlcom (VaR95: _b[mu:_cons] + sqrt(_b[v:_cons])*invt(5,0.05))
> (VaR97: _b[mu:_cons] + sqrt(_b[v:_cons])*invt(5,0.03))
> (VaR99: _b[mu:_cons] + sqrt(_b[v:_cons])*invt(5,0.01)), noheader
```

	Coef.	Std. Err.	z	P>|z|	[95% Conf.	Interval]
VaR95	-.022974	.0005387	-42.65	0.000	-.0240298	-.0219181
VaR97	-.0276672	.0006374	-43.41	0.000	-.0289165	-.0264178
VaR99	-.0385575	.0008695	-44.34	0.000	-.0402616	-.0368533

We compute the inverse of a Student's t distribution by using the `invt()` function, taking as the first argument the number of degrees of freedom and as the second argument the percentile. To obtain a fat-tail distribution, we specify 5 degrees of freedom. The VaR under the assumption of a Student's t distribution is by far higher, in absolute terms, than those obtained under the normality case, reflecting the fatter tails of the Student's t distribution compared with the normal one. The main differences are observed in correspondence to the highest quantiles.

One of the main issues when estimating VaR using the parametric approach is the estimation of the parameters, mainly with reference to volatilities and, when considering a portfolio of assets, to the correlation matrix. To estimate volatilities, a solution is to use historical volatilities, simply computed as the standard deviation of past returns, as illustrated above. An alternative is to get some volatility forecast, for instance, by fitting a GARCH model on historical data and afterward computing a volatility forecast at one period ahead. Another opportunity is to adopt a forward-looking setting by computing the implied volatility in the Black–Scholes framework. However, the Black–Scholes model is imperfect because it assumes that the volatility σ is constant, while implied volatility depends on option characteristics, for instance, on time to expiration and on strike price. To compute the correlation matrix, we could also either use historical data or adopt some more sophisticated estimators, such as the correlation matrix, that we can obtain from an MGARCH model, for instance, a DCC model.

We now show how to use GARCH models to compute VaR in a univariate setting. We start by fitting a standard GARCH(1,1) on the S&P 500 returns, and then we use the obtained volatility to estimate VaR referring to (5.4).

```
arch SP500, arch(1) garch(1)
predict variance, variance
generate VaRGARCH95 = mean+variance^0.5*invnormal(0.05)
generate VaRGARCH97 = mean+variance^0.5*invnormal(0.03)
generate VaRGARCH99 = mean+variance^0.5*invnormal(0.01)
```

We now represent the historical returns of the S&P 500 with the estimated VaR at the three confidence levels:

```
. tsset date
        time variable:  date, 1/2/1991 to 12/30/2013, but with gaps
                delta:  1 day
. tsline SP500 VaRGARCH95 VaRGARCH97 VaRGARCH99
```

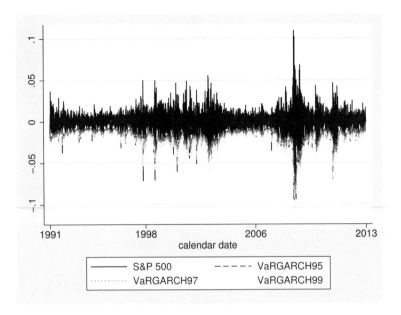

Figure 5.1. VaR–GARCH normal distribution

The VaR at the three confidence levels form a lower bound to the daily returns. To evaluate the goodness of this model in predicting VaR, we adopt formal tests that we describe in an upcoming section. The hypotheses that we are going to test are whether the percentage of violations—the times returns exceed the VaR—is close enough to the confidence level at which we compute the VaR and whether these violations occur independently from each other.

By using the following commands, we can compute the standard errors of the VaR we obtained:

```
. tsset newdate
        time variable:  newdate, 02jan1991 to 30dec2013
                delta:  1 day
. predict residuals, residuals
. predictnl VaR = mean+([ARCH]_b[_cons]+ [ARCH]_b[L.arch]*L.residuals^2+
> [ARCH]_b[L.garch]*L.variance)^0.5*invnormal(0.05), se(VaR_SE)
(1 missing value generated)
```

```
. generate VaRPlusSE = VaR+VaR_SE
(1 missing value generated)
. generate VaRMinusSE = VaR-VaR_SE
(1 missing value generated)
```

First, we computed residuals from the GARCH model by using the `predict` command and specifying the `residuals` option. After that, we use the `predictnl` command to obtain standard errors by specifying the `se` option followed by the name of the variable in which we want to store standard errors, VaR_SE. Note that to obtain standard errors, we use the parameters estimated by GARCH to compute the variance, namely, `[ARCH]_b[_cons]`, `[ARCH]_b[L.arch]`, and `[ARCH]_b[L.garch]`.

We represent the VaR with the upper and lower standard errors (see figure 5.2).

```
. tsset date
        time variable:  date, 1/2/1991 to 12/30/2013, but with gaps
                delta:  1 day
. tsline SP500 VaR VaRPlusSE VaRMinusSE
```

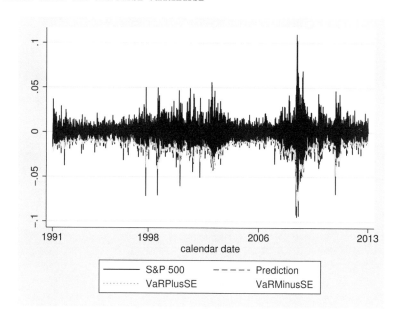

Figure 5.2. VaR–GARCH normal standard errors

Because the standard errors are very small, it is almost impossible to distinguish VaR from the upper and lower bounds.

We now evaluate a GARCH under the assumption that returns are distributed as a Student's t with 5 degrees of freedom, both when estimating VaR as well as when computing the critical value.

```
tsset newdate
drop variance
arch SP500, arch(1) garch(1) distribution(t) nolog
predict variance, variance
generate VaRGARCHT95 = mean+variance^0.5*invt(5,0.05)
generate VaRGARCHT97 = mean+variance^0.5*invt(5,0.03)
generate VaRGARCHT99 = mean+variance^0.5*invt(5,0.01)
```

To get some preliminary ideas about the alternative fit of the two models in computing VaR, we report figure 5.3, where we compare VaR at the 95% confidence level estimated under both the normal and the Student's t distribution.

```
. tsset date
        time variable:  date, 1/2/1991 to 12/30/2013, but with gaps
                delta:  1 day

. tsline SP500 VaRGARCH95 VaRGARCHT95
```

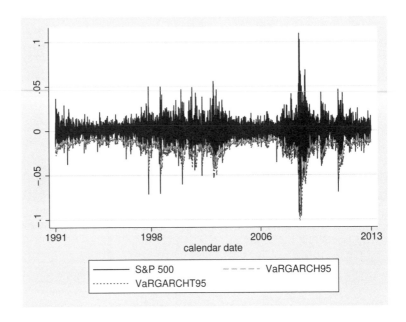

Figure 5.3. VaR–GARCH normal and Student's t distribution

Figure 5.3 shows that the Student's t distribution implies a lower VaR than the normal distribution does. To determine which model should be preferred, we need to use a formal test that allows us to verify which model implies an average number of violations close to the confidence level.

If we are interested in computing the VaR not for a single position but for a portfolio, we have to think about the correlation between the assets composing the portfolio. For instance, considering an equally weighted portfolio formed by only two assets, X_1 and X_2, both i.i.d., we can compute the covariance of returns as

$$\sigma_{12}^2 = \rho_{12}\sigma_1\sigma_2 \tag{5.5}$$

where σ_1 and σ_2 are the volatilities (standard deviations) of X_1 and X_2, respectively, and ρ_{12} is the correlation between assets X_1 and X_2. The square root of (5.5) is the volatility of the portfolio that we can use to compute the VaR of the portfolio by plugging this quantity into (5.4).

Under the assumption that the two positions are normally distributed, the VaR of the portfolio will also follow a normal distribution because a linear combination of a normal variable is still a normal variable. Alternatively, given the VaR of the two assets, VaR_1 and VaR_2 with correlation ρ_{12}, we can compute the VaR of the portfolio VaR_p as

$$\mathrm{VaR}_p = \sqrt{\rho_{1,2}\mathrm{VaR}_1\mathrm{VaR}_2} \tag{5.6}$$

We now consider an equally weighted portfolio of two assets, the S&P 500 and the FTSE 100, observed over 1991–2013 at daily frequency. To compute the VaR associated with the portfolio, we need first to estimate the following parameters with the GMM method: mean and variance for S&P 500, mean and variance for FTSE 100, and correlation between S&P 500 and FTSE 100.

```
. quietly summarize SP500
. local n = r(N)
. matrix A = (r(mean), r(Var))
. quietly summarize FTSE100
. matrix B = (r(mean), r(Var))
. quietly correlate SP500 FTSE100
. matrix init = A, B, r(rho)
```

```
. gmm (SP500-{mu_SP500})
> ((SP500-{mu_SP500})^2 - (`n´)/(`n´-1)*{v_SP500})
> (FTSE100-{mu_FTSE100})
> ((FTSE100-{mu_FTSE100})^2 - (`n´)/(`n´-1)*{v_FTSE100})
> ((SP500-{mu_SP500})*(FTSE100-{mu_FTSE100})/(sqrt({v_SP500})*
> sqrt({v_FTSE100}))-{rho}),
> onestep winitial(identity) vce(robust) from(init)

Step 1
Iteration 0:   GMM criterion Q(b) =  7.212e-09
Iteration 1:   GMM criterion Q(b) =  3.397e-15

note: model is exactly identified

GMM estimation

Number of parameters =    5
Number of moments    =    5
Initial weight matrix: Identity              Number of obs   =      5,828
```

	Coef.	Robust Std. Err.	z	P>|z|	[95% Conf. Interval]	
/mu_SP500	.0002885	.0001512	1.91	0.056	-7.89e-06	.0005848
/v_SP500	.0001332	5.80e-06	22.98	0.000	.0001219	.0001446
/mu_FTSE100	.0001891	.0001484	1.27	0.203	-.0001019	.00048
/v_FTSE100	.0001284	4.82e-06	26.62	0.000	.000119	.0001379
/rho	.4950217	.018541	26.70	0.000	.458682	.5313615

```
Instruments for equation 1: _cons
Instruments for equation 2: _cons
Instruments for equation 3: _cons
Instruments for equation 4: _cons
Instruments for equation 5: _cons
```

Now we can compute the VaR portfolio at 95%, 97%, and 99% confidence levels with their confidence intervals (5.6):

```
. nlcom (VaRP: -sqrt((_b[mu_SP500:_cons] + sqrt(_b[v_SP500:_cons]))*
> (_b[mu_FTSE100:_cons] + sqrt(_b[v_FTSE100:_cons]))*
> invnormal(0.95)*_b[rho:_cons]))

        VaRP:  -sqrt((_b[mu_SP500:_cons] + sqrt(_b[v_SP500:_cons]))*
> (_b[mu_FTSE100:_cons] + sqrt(_b[v_FTSE100:_cons]))*
> invnormal(0.95)*_b[rho:_cons])
```

	Coef.	Std. Err.	z	P>|z|	[95% Conf. Interval]	
VaRP	-.0105348	.000327	-32.22	0.000	-.0111756	-.0098939

```
. nlcom (VaRP: -sqrt((_b[mu_SP500:_cons] + sqrt(_b[v_SP500:_cons]))*
> (_b[mu_FTSE100:_cons] + sqrt(_b[v_FTSE100:_cons]))*
> invnormal(0.97)*_b[rho:_cons]))

        VaRP:  -sqrt((_b[mu_SP500:_cons] + sqrt(_b[v_SP500:_cons]))*
> (_b[mu_FTSE100:_cons] + sqrt(_b[v_FTSE100:_cons]))*
> invnormal(0.97)*_b[rho:_cons])
```

	Coef.	Std. Err.	z	P>|z|	[95% Conf. Interval]	
VaRP	-.011265	.0003497	-32.22	0.000	-.0119503	-.0105797

```
. nlcom (VaRP: -sqrt((_b[mu_SP500:_cons] + sqrt(_b[v_SP500:_cons]))*
> (_b[mu_FTSE100:_cons] + sqrt(_b[v_FTSE100:_cons]))*
> invnormal(0.99)*_b[rho:_cons]))

        VaRP:  -sqrt((_b[mu_SP500:_cons] + sqrt(_b[v_SP500:_cons]))*
> (_b[mu_FTSE100:_cons] + sqrt(_b[v_FTSE100:_cons]))*
> invnormal(0.99)*_b[rho:_cons])
```

	Coef.	Std. Err.	z	P>|z|	[95% Conf. Interval]	
VaRP	-.0125285	.0003889	-32.22	0.000	-.0132906	-.0117663

The VaR of the portfolio at the 95% confidence level at 1 day is -1.05% with a confidence level of $[-1.12\%, -0.99\%]$. The VaR of the portfolio is lower than the VaR associated with each single position thanks to the diversification effect. In fact, the two assets are just mildly correlated with a correlation coefficient equal to 0.50.

Similarly to when we obtained a time series of VaR by using the volatility computed by a univariate GARCH model, we now compute the VaR for the portfolio by using as input a dynamic time series of correlations obtained from a DCC model. We start by fitting a DCCE model for the S&P 500 and FTSE 100 under the assumption that returns are distributed as a multivariate normal.

```
. tsset newdate
        time variable:  newdate, 02jan1991 to 30dec2013
                delta:  1 day

. mgarch dcc SP500 FTSE100, arch(1) garch(1) nolog

Dynamic conditional correlation MGARCH model

Sample: 02jan1991 - 30dec2013                  Number of obs   =       5,828
Distribution: Gaussian                         Wald chi2(.)    =          .
Log likelihood =  38660.85                     Prob > chi2     =          .
```

	Coef.	Std. Err.	z	P>|z|	[95% Conf.	Interval]
SP500						
_cons	.0006388	.0001016	6.29	0.000	.0004396	.000838
ARCH_SP500						
arch						
L1.	.0760627	.0061513	12.37	0.000	.0640063	.088119
garch						
L1.	.913811	.0067377	135.63	0.000	.9006053	.9270167
_cons	1.25e-06	1.96e-07	6.35	0.000	8.62e-07	1.63e-06
FTSE100						
_cons	.0005054	.0001064	4.75	0.000	.0002968	.000714
ARCH_FTSE100						
arch						
L1.	.0792091	.0063074	12.56	0.000	.0668468	.0915714
garch						
L1.	.9096009	.0070447	129.12	0.000	.8957936	.9234082
_cons	1.38e-06	2.36e-07	5.86	0.000	9.21e-07	1.85e-06
corr(SP500,						
FTSE100)	.5056933	.0724493	6.98	0.000	.3636952	.6476913
Adjustment						
lambda1	.0058031	.0009408	6.17	0.000	.0039593	.007647
lambda2	.9931893	.0010976	904.84	0.000	.9910379	.9953406

Once we have fit the DCCE model, we want to obtain the in-sample estimates of
the covariance between the two return time series and plug its square root into (5.4) to
compute the VaR.

```
. predict variance*, variance
. quietly summarize SP500
. generate mean1 = r(mean)
. quietly summarize FTSE100
. generate mean2 = r(mean)
. generate meanP = mean1*0.5+mean2*0.5
. generate VaRPDCC95 = meanP+sqrt(variance_FTSE100_SP500)*invnormal(0.05)
. generate VaRPDCC97 = meanP+sqrt(variance_FTSE100_SP500)*invnormal(0.03)
. generate VaRPDCC99 = meanP+sqrt(variance_FTSE100_SP500)*invnormal(0.01)
```

Finally, we compute the time series of portfolio returns and plot it with the VaR estimated at the 95%, 97%, and 99% confidence levels (see figure 5.4):

```
. generate returnP = SP500*0.5+FTSE100*0.5
. tsset date
        time variable:  date, 1/2/1991 to 12/30/2013, but with gaps
                delta:  1 day
. tsline returnP VaRPDCC95 VaRPDCC97 VaRPDCC99
```

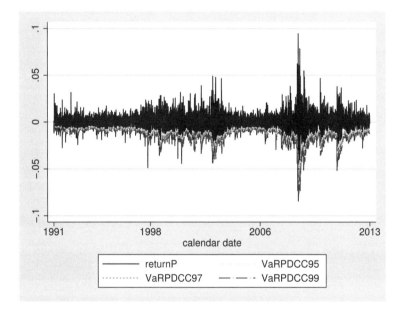

Figure 5.4. VaR–DCC normal

We can proceed in a similar way under the assumption that innovations follow a multivariate Student's t distribution (see figure 5.5):

```
. tsset newdate
        time variable:  newdate, 02jan1991 to 30dec2013
                delta:  1 day
. mgarch dcc SP500 FTSE100, arch(1) garch(1) distribution(t) nolog
Dynamic conditional correlation MGARCH model
Sample: 02jan1991 - 30dec2013                Number of obs   =       5,828
Distribution: t                              Wald chi2(.)    =           .
Log likelihood =  38849.58                   Prob > chi2     =           .
```

	Coef.	Std. Err.	z	P>|z|	[95% Conf. Interval]	
SP500						
_cons	.0007049	.0000961	7.34	0.000	.0005166	.0008932
ARCH_SP500						
arch						
L1.	.0672798	.0063085	10.66	0.000	.0549154	.0796443
garch						
L1.	.9251138	.0067028	138.02	0.000	.9119765	.9382511
_cons	8.69e-07	1.80e-07	4.84	0.000	5.17e-07	1.22e-06
FTSE100						
_cons	.0005664	.0001032	5.49	0.000	.0003642	.0007687
ARCH_FTSE100						
arch						
L1.	.0712458	.0067597	10.54	0.000	.0579971	.0844945
garch						
L1.	.9194466	.0073731	124.70	0.000	.9049956	.9338976
_cons	1.23e-06	2.44e-07	5.04	0.000	7.53e-07	1.71e-06
corr(SP500, FTSE100)	.4648721	.099296	4.68	0.000	.2702555	.6594886
Adjustment						
lambda1	.0073205	.0014253	5.14	0.000	.004527	.0101139
lambda2	.9916408	.0016701	593.77	0.000	.9883675	.9949141
df						
_cons	8.451905	.5956692	14.19	0.000	7.284415	9.619395

```
. predict varianceT*, variance
. generate VaRPDCCT95 = meanP+sqrt(varianceT_FTSE100_SP500)*invnormal(0.05)
. generate VaRPDCCT97 = meanP+sqrt(varianceT_FTSE100_SP500)*invnormal(0.03)
. generate VaRPDCCT99 = meanP+sqrt(varianceT_FTSE100_SP500)*invnormal(0.01)
. tsset date
        time variable:  date, 1/2/1991 to 12/30/2013, but with gaps
                delta:  1 day
. tsline returnP VaRPDCCT95 VaRPDCCT97 VaRPDCCT99
```

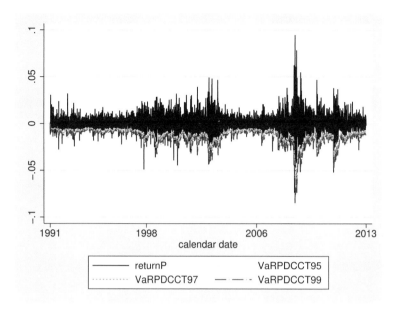

Figure 5.5. VaR–DCC Student's t

In this example, we obtained the percentile adopting the normal distribution, but we could also use the Student's t distribution.

The main advantage of the parametric approach is that it relies on a simple framework such that we can compute VaR in a straightforward way. The counterpart is that the distribution assumed to compute the percentile should properly fit the loss distribution. For instance, if we decide to adopt the normal distribution and the empirical distribution is indeed leptokurtic, we will underestimate the tail-risk. On the other side, when choosing a fat-tail distribution while the empirical distribution is not, we could get an overestimate of the VaR. We need to understand that even this second case can be painful for a financial institution, because it could be forced to use too much capital to meet safety regulations and therefore incur opportunity costs from the missed opportunity to invest that money in an alternative way. It is thus important to properly select the distribution describing loss to avoid both over- and underestimation of risk. The VaR backtesting procedures we are going to illustrate later in the chapter will help us on that point.

A final drawback of the parametric approach is that the assumption about a linear relationship between risk factors and financial assets may not always offer a good approximation of the true relationship, especially when computing the risk of assets with nonlinear pay-offs, like options.

5.4.3   Historical simulation

The historical simulation approach assumes that we can draw risk factors from historical ones. In a nutshell, the estimation of VaR occurs in two steps. In the first step, we compute returns over the horizon time τ for each asset in the portfolio. In the second step, we apply historical returns to the most recent value of the portfolio, to compute a distribution of future portfolio values. We finally obtain VaR as the percentile of the simulated loss distribution at the confidence level α we are interested in.

More formally, let us consider the historical loss distribution:

$$\widetilde{L}_t = l_t\left(X_s\right) \qquad s = T - N + 1, \ldots, T$$

with $\widetilde{L}_t$ indicating what would happen to the current portfolio if the past risk factor changes on day s were to recur. T is the length of the time series, the overall number of possible values to be used for the computation of the historical simulated portfolio values, and N is the number of values effectively used.

Having obtained a distribution of possible future portfolio values, we compute the VaR as the αth quantile. Let us consider the N ordered possible variations of portfolio values $\widetilde{L}_{N,N} \leq \cdots \leq \widetilde{L}_{1,N}$. We define VaR as

$$\mathrm{VaR}_\alpha(L) = \widetilde{L}_{\{N(1-\alpha)\},N}$$

where $N(1-\alpha)$ denotes the largest integer not exceeding $N(1-\alpha)$, the $(1-\alpha)$th percentile.

Let us go back to the historical daily returns of the S&P 500 for the years 1999–2013 and estimate the VaR in the historical simulation framework. We consider the historical returns that we represent in the following histogram (see figure 5.6) to be possible tomorrow changes in the stock index:

```
. histogram SP500
(bin=37, start=-.09469514, width=.00552073)
```

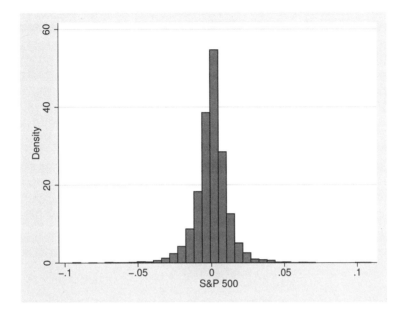

Figure 5.6. S&P historical returns

We obtain VaR as the αth percentile of the distribution in figure 5.6, which we can compute as follows:

```
. centile SP500, centile(1,3,5)
```

				— Binom. Interp. —	
Variable	Obs	Percentile	Centile	[95% Conf. Interval]	
SP500	5,828	1	-.0321395	-.0352479	-.0300883
		3	-.0224311	-.0237288	-.0210248
		5	-.0176973	-.0185306	-.0167888

The `centile` command allows us to obtain not only the percentiles we are interested in but also their confidence intervals. For instance, in this example, we can see that the VaR at the 95% confidence level at 1 day is -3.21% with a confidence interval of $[-3.52\%, 3.01\%]$. We could even decide to limit the sample of possible scenarios, excluding, for instance, the oldest returns recorded before year 2000:

```
. centile SP500 if date >= mdy(01,01,2000), centile(1,3,5)
```

				— Binom. Interp. —	
Variable	Obs	Percentile	Centile	[95% Conf.	Interval]
SP500	3,571	1	-.0362922	-.0437211	-.0332708
		3	-.025106	-.0272434	-.023869
		5	-.0205206	-.0222895	-.0189797

Interestingly, when we exclude data before the year 2000, we come up with higher absolute values, reflecting the lower volatility characterizing the 1990s. Moreover, we can see that VaRs computed in the historical simulation framework are higher than those obtained following the parametric approach under the assumption of normal distribution, while they are lower than those following the parametric approach with Student's t.

When working with a portfolio of assets, we can proceed in a similar way. We first have to generate the historical distribution of portfolio returns, to compute afterward its empirical percentiles. For instance, consider again an equally weighted portfolio composed of the S&P 500 and FTSE 100. We can get the historical distribution of portfolio returns as follows (see figure 5.7):

```
. generate returnportfolio = SP500*0.5+FTSE100*0.5

. histogram returnportfolio, normal width(0.001)
(bin=179, start=-.08449084, width=.001)
```

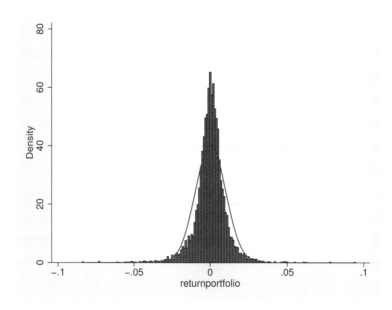

Figure 5.7. S&P 500 and FTSE 100 portfolio historical returns

In the `histogram` command, we specified the `normal` option to compare the normal with the empirical distribution of portfolio returns. This chart shows that the two distributions are far from being equal.

Having generated a set of possible scenarios, we can compute the VaR for the portfolio as we have done before:

```
. _pctile returnportfolio, p(1,3,5)
. return list
scalars:
                r(r1) =  -.028185972943902
                r(r2) =  -.0192605629563332
                r(r3) =  -.0156204840168357
```

The historical simulation approach offers a simple framework with the great advantage of not imposing any distribution assumptions on P&L. Contrary to what happens for the parametric approach, the historical simulation framework allows for a nonlinear dependence between assets in the portfolio and the underlying risk factors. Finally, another advantage is that when considering a portfolio, we do not need to care about the dependence structure of assets within the portfolio because it is already reflected in historical returns: if two assets are positively correlated, then they will tend to show returns with the same sign and vice versa.

Aside from these advantages that the historical simulation offers compared with the parametric approach, the main limitation of the historical simulation approach is that it relies on the assumption that the future replicates the past: the possible future realizations of price processes are only those that already occurred in the past. This may underestimate the occurrence of extreme scenarios when no extreme event is included in the historical data from which we are drawing. For instance, thinking about Black Monday, if this day is not included in the time series we are considering, then the probability that it occurs again in the future is equal to 0. Instead, when using a parametric design approach, we just need to adopt a proper tail distribution to account for possible extreme events.

5.4.4   Monte Carlo simulation

Finally, we present the Monte Carlo simulation methodology that is widely used in the financial industry. The first step requires us to choose a model that could reasonably describe the time evolution of risk factors $\widetilde{X}_{t+1} = X_{t-N+1}, \ldots, X_t$. For instance, to simulate the possible future scenarios of equity assets, we could assume a Brownian motion whose parameters need to be estimated. Once we have defined the model for risk factors, we can start by generating a number m of independent realizations of risk-factor changes for the next time period $t+1$: $X_{t+1}^{(1)}, \ldots, X_{t+1}^{(N)}$. The possible $i = 1, \ldots, N$ scenarios are the input of the pricing formulas that we have to use to obtain the simulated P&L distribution:

$$\widetilde{L}_{t+1}^{(i)} = l_{[t]} \left\{ \widetilde{X}_{t+1}^{(i)} \right\}$$

Now that we have generated a simulated loss distribution, we can obtain VaR by simply computing the $(1 - \alpha)$th percentile. Formally, let $\widetilde{L}_{N,N} \leq \cdots \leq \widetilde{L}_{1,N}$ be N ordered possible losses. We can compute VaR as

$$\mathrm{VaR}_\alpha(L) = \widetilde{L}_{\{N(1-\alpha)\},N}$$

where $N(1 - \alpha)$ denotes the largest integer not exceeding $N(1 - \alpha)$.

As an empirical example, consider again the daily returns of the S&P 500 over the years 1999–2013. Generate many possible scenarios, say, 10,000, assuming returns can be properly described by a normal distribution with a mean and variance equal to their empirical historical counterparts. We start by defining the length of the simulation with the **set obs** command. The next step requires us to estimate the historical mean and variance and to simulate the normal random numbers with the just estimated parameters by using the **rnormal()** function.

```
. use http://www.stata-press.com/data/feus/index, clear
. set obs 10000
number of observations (_N) was 5,828, now 10,000
. quietly summarize SP500
. scalar mean = r(mean)
. scalar std = r(sd)
. set seed 1
. generate simulatedreturns = rnormal(mean, std)
```

Using the `histogram` command, we can obtain a graphical representation of the simulated returns (see figure 5.8):

```
. histogram simulatedreturns
(bin=40, start=-.04712843, width=.00234801)
```

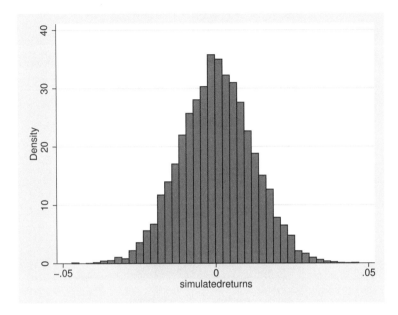

Figure 5.8. S&P simulated returns—normal

Once we have obtained the distribution of simulated returns, we can compute VaR as the $(1 - \alpha)$th percentile:

```
. centile simulatedreturns, centile(1,3,5)
```

| | | | | — Binom. Interp. — | |
Variable	Obs	Percentile	Centile	[95% Conf. Interval]
simulatedr~s	10,000	1	-.0272083	-.0280109	-.0262179
		3	-.021969	-.0224987	-.0214302
		5	-.0189585	-.0193896	-.018546

The VaR we have just obtained are quite close to those we got using the parametric method under the assumption of normality; in fact, the mean and standard deviation are exactly the same as those we got in that case.

Of course, we could assume an alternative distribution to the normal one, for instance, the Student's t. In this case, we have to use the `rt()` function taking as an argument the number of degrees of freedom that we set equal to 5. Because the function does not allow for a location-scale family, we will adjust afterward the numbers extracted from a Student's t with the empirical mean and standard deviation:

```
. set seed 1
. generate simulatedT = rt(5)
. generate simulatedT_LocationScale = simulatedT*std+mean
. histogram simulatedT_LocationScale
(bin=40, start=-.14921091, width=.00635544)
```

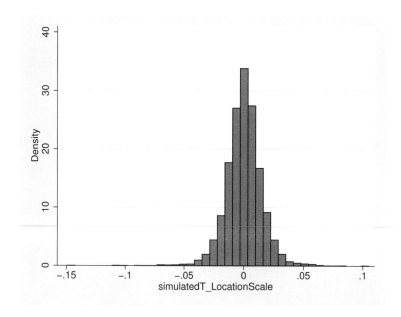

Figure 5.9. S&P simulated returns— Student's $t(5)$

The distribution of returns simulated according to a Student's t distribution with 5 degrees of freedom depicted in figure 5.9 is characterized by much fatter tails than those we got under the assumption of Gaussianity and reported in figure 5.8. Therefore, we expect higher VaR when using the Student's t than when adopting the normal distribution. This is exactly what we find:

```
. centile simulatedT_LocationScale, centile(1,3,5)
```

| | | | | — Binom. Interp. — | |
Variable	Obs	Percentile	Centile	[95% Conf. Interval]	
simulatedT~e	10,000	1	-.0370054	-.0393441	-.0353573
		3	-.0263742	-.027884	-.0256998
		5	-.0225608	-.023466	-.0219087

Comparing those percentiles with those obtained previously under the assumption of normal distribution, we can note that these ones are lower, with the difference between the two distributions vanishing as the confidence level decreases. This result is coherent with the Student's t being characterized by fatter tails with respect to the normal.

When moving from a single asset to a portfolio of assets, and with the purpose of computing VaR, we should be able to generate scenarios drawing from a multivariate distribution. For instance, assuming a bivariate normal distribution, we can start by estimating historical parameters, namely means, volatilities and correlations, and afterward we can use these quantities to initialize the drawings. The command allowing us to generate simulated multivariate normal numbers is `drawnorm`.

We consider again the case of an equally weighted portfolio composed by the S&P 500 and FTSE 100 and compute their means, volatilities, and correlations by the usual `summarize` command. Because the `drawnorm` command requires matrices as input, we should store all the historical statistics in matrices by using the `matrix` command rather than the usual `generate` command.

```
. summarize SP500
    Variable │        Obs        Mean    Std. Dev.        Min        Max
─────────────┼──────────────────────────────────────────────────────────
       SP500 │      5,828    .0002885    .0115444   -.0946951    .109572
. matrix mean1 = r(mean)
. matrix sd1 = r(sd)
. summarize FTSE100
    Variable │        Obs        Mean    Std. Dev.        Min        Max
─────────────┼──────────────────────────────────────────────────────────
     FTSE100 │      5,828    .0001891    .0113336   -.0926456   .0938424
. matrix mean2 = r(mean)
. matrix sd2 = r(sd)
. matrix meanV = (mean1,mean2)
. matrix sdV = (sd1,sd2)
. correlate SP500 FTSE100
(obs=5,828)
             │    SP500  FTSE100
─────────────┼──────────────────
       SP500 │   1.0000
     FTSE100 │   0.4949   1.0000
. matrix correlation = r(C)
```

In the first part of the code, we obtain the mean and standard deviation for both the S&P 500 and the FTSE 100 that we stored in matrices, `mean1` and `sd1` for S&P 500, and `mean2` and `sd2` for FTSE 100. Then, we create two vectors, one for means (`meanV`) and one for standard deviations (`sdV`), putting together the parameters for the two time series. Finally, we compute the historic correlation and store that value in the matrix `correlation`.

Once we have stored parameters, we can draw simulations as follows:

```
. set seed 1
. drawnorm SPSimulated FTSESimulated, n(10000) means(meanV) sds(sdV)
> corr(correlation)
```

Using `drawnorm`, we have generated 10,000 pairs of random numbers extracted from a bivariate normal with the previously stored parameters, `meanV`, `sdV`, and `correlation`. The new variables are labeled `SPSimulated` and `FTSESimulated`.

To check whether the simulated values exhibit properties in line with parameters used for simulation, we compute means, standard deviations, and correlation of the two simulated series.

```
. summarize SPSimulated FTSESimulated

    Variable |        Obs        Mean    Std. Dev.        Min         Max

  SPSimulated |     10,000    .0002787    .0117951   -.0471284    .0467921
 FTSESimula~d |     10,000    .0001227    .0115599   -.0421071    .0414804

. correlate SPSimulated FTSESimulated
(obs=10,000)

             | SPSimu~d FTSESi~d

  SPSimulated |  1.0000
 FTSESimula~d |  0.5029   1.0000
```

From this quick check, it is evident that the simulated parameters are quite close to the historic ones we used as input in our simulation process.

The simulations we have in hand are for one period and for each stock index considered. Our final goal is to get a VaR of an equally weighted portfolio composed by the two assets; therefore, we compute the simulated portfolio distribution as a simple weighted average of the simulated returns, which we represent in figure 5.10.

```
. generate simulateP = SPSimulated*0.5+FTSESimulated*0.5

. histogram simulateP, normal width(0.001)
(bin=79, start=-.03897972, width=.001)
```

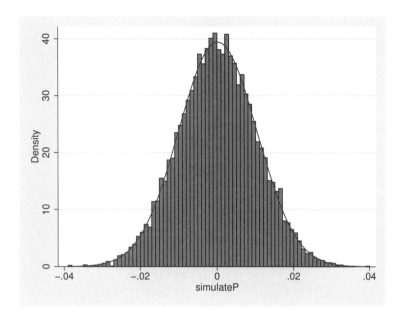

Figure 5.10. S&P 500 and FTSE 100 simulated returns—bivariate normal

From figure 5.10, we can see that the simulated portfolio returns are close to a normal distribution. In fact, we have generated returns under the assumption of multivariate normal distribution.

The last step consists of computing the percentiles of this distribution at the confidence levels in which we are interested:

```
. _pctile(simulateP), p(1 3 5)

. return list

scalars:
                r(r1) =  -.0233444031327963
                r(r2) =  -.0188167728483677
                r(r3) =  -.0164427533745766
```

Comparing these values with those obtained using the historical simulation approach, we note that the difference between the two methods gets bigger as the confidence level increases. This result suggests that most likely we are underestimating the tail risk in this Monte Carlo simulation. This comes not as a surprise given that we are using a multivariate normal distribution that is not a fat-tail distribution. If we had chosen a multivariate Student's t distribution, we would have obtained lower percentiles.

The great advantage of the Monte Carlo approach is its extreme flexibility, because we can specify all the models from which we wish to draw possible scenarios for risk factors. Of course, this implies that we accept the so-called model risk, which is the risk of assuming a wrong model for describing and simulating scenarios. For instance, if we assume that we can reasonably describe future returns by using a normal distribution ignoring fat tails, we could underestimate the VaR. This limitation affects even the parametric approach. Note that the Monte Carlo simulation approach can be quite time consuming, especially if the portfolio includes several financial derivatives that we cannot price using a closed form.

5.4.5   Expected shortfall

Despite its great success and wide application in the financial industry, VaR suffers from noticeable limitations, the most severe being that it does not provide any information about the loss we could experience once the limit identified by VaR has been passed.

The expected shortfall (ES) was explicitly designed to overcome that limitation. ES addresses the following question:

> If I'm going to experience a loss, how severe will this loss be?

In this respect, ES is a more conservative measure than VaR because it looks at the average of all losses exceeding the VaR.

Let L be the loss with $E(|L|) < \infty$ and probability density function (p.d.f.) f_L. We can define ES at $\alpha \in (0,1)$ confidence level as

$$\mathrm{ES}_\alpha = \frac{1}{1-\alpha} \int_\alpha^1 \mathrm{VaR}_u\left(f_L\right) du$$

From this definition, we can see that ES is the weighted average of losses lying on the right side of the VaR.

Under the assumption that the loss distribution L follows a normal distribution with mean μ and variance σ^2, the ES takes the form

$$\mathrm{ES}_\alpha = \mu + \sigma E\left\{\frac{L-\mu}{\sigma} \left| \frac{L-\mu}{\sigma} \geq q_\alpha\left(\frac{L-\mu}{\sigma}\right)\right.\right\} \tag{5.7}$$

where q_α is the αth percentile of the standardized loss distribution. We can easily estimate (5.7) as

$$\mathrm{ES}_\alpha = \mu + \sigma \frac{\phi\left\{\Phi^{-1}(1-\alpha)\right\}}{1-\alpha}$$

where Φ and ϕ are the c.d.f. and p.d.f. of a normal, respectively, while μ and σ are the sample mean and standard deviation, respectively.

We now compute the ES under the assumption of normal distribution with three confidence levels, 95%, 97%, and 99%:

```
. scalar ES95 = normalden(invnormal(1-0.95))/(1-0.95)
. scalar ES97 = normalden(invnormal(1-0.97))/(1-0.97)
. scalar ES99 = normalden(invnormal(1-0.99))/(1-0.99)
. display ES95,ES97,ES99
2.0627128 2.268065 2.6652142
```

We can read these values as follows: once the VaR threshold has been broken, we can incur a loss of 2.06% at the 95% confidence level, of 2.27% at the 97%, and of 2.67% at the 99%.

From a methodological point of view, the great advantage of ES over VaR is that it is a coherent risk measure given that it satisfies even the subadditivity property. Given two positions on assets 1 and 2, we could always verify that $ES(L_1 + L_2) \leq ES(L_1) + ES(L_2)$, while the same property does not hold true for VaR. Additionally, ES puts fewer restrictions on the loss distribution because it requires only the first moment to be finite, while in the case of VaR, we need even the second central moment to be finite.

5.5   Backtesting procedures

So far, we have presented some alternative methods to estimate VaR and ES, highlighting pros and cons of each method. In what follows, we introduce some backtesting procedures that we use to properly compare the alternative approaches adopted to compute VaR and ES.

Backtesting procedures are constructed by comparing realized returns and model-generated VaR measures, most of them relying on the analysis of the hit sequence or VaR violations that we can define as

$$I_t = \begin{cases} 1 & \text{if } L_t < -\text{VaR}_t(\alpha) \\ 0 & \text{otherwise} \end{cases} \tag{5.8}$$

The hit sequence is a sequence of dummy variables taking a value of 1 when we experience a higher loss than that identified by the VaR and a value of 0 otherwise.

Commonly used VaR backtesting procedures include the likelihood-ratio test by Kupiec (1995), the Markov tests by Christoffersen (1998), and the duration-based test by Christoffersen and Pelletier (2004).

We can classify the tests for assessing the goodness of a risk measure into three categories:

1. unilevel VaR tests

2. multilevel VaR tests

3. expected shortfall and tail-risk tests

The unilevel backtesting procedures are so called because they focus separately on alternative quantiles. If we want to assess how good a specific methodology is at estimating VaR at alternative confidence levels—say, 95%, 97%, and 99%—then we should carry out a separate test for each alternative. With this approach, especially when we are considering high confidence levels, we can have just a small number of observations at our disposal, decreasing the power of the test.

Recently, multilevel procedures have been introduced explicitly to overcome this limit; in fact, they allow us to assess the performance of a specific approach at measuring VaR by considering alternative coverage levels simultaneously. This kind of test shows good performance in terms of power and also prevents banks and other financial institutions from providing information on the shape of their tail losses. The unwillingness of banks to disclose information beyond one-day-ahead VaR estimates is the main drawback of tail-risk tests. Indeed, this last category of backtesting procedures does not focus on one or more percentiles but on the entire shape of the profit and loss distribution.

In the remainder of the chapter, we focus only on unilevel procedures because they are the most common tests adopted by financial institutions. For the interested reader, we provide some references that focus on the other two categories: for multilevel tests, see Perignon and Smith (2008), Hurlin and Tokpavi (2006), and Leccadito, Boffelli, and Urga (2014); for tail-risk tests, see Berkowitz and O'Brien (2002).

5.5.1   Unilevel VaR tests

The most famous backtesting procedures for the VaR are the unconditional coverage (uc), the independence (ind), and the conditional coverage (cc) tests, all introduced by Christoffersen (1998), and the Weibull duration test, introduced by Christoffersen and Pelletier (Christoffersen and Pelletier (2004)).

The unconditional coverage test

In the unconditional coverage (uc) test, the null hypothesis under investigation is that the sequence I_t in (5.8) is i.i.d. Bernoulli with parameter $(1 - \alpha)$ against the alternative that the Bernoulli parameter is $\widehat{\pi}_1$, where $\widehat{\pi}_1$ is the empirical ratio of violations $T1/T$, where $T1$ is the number of days with violations. If the VaR method is correct, then the empirical failure rate $\widehat{\pi}$ must be equal to $(1 - \alpha)$. The likelihood function of a Bernoulli variable z with parameter p is

$$L(z; p) = (1 - p)^{T - T1} p^{T1} \tag{5.9}$$

and therefore we construct the likelihood-ratio test for the uc as

$$\text{LR}_{\text{uc}} = 2 \left\{ \ln L(z; \widehat{\pi}_1) - \ln L(z; p) \right\} \sim \chi_1^2 \tag{5.10}$$

We now check whether the VaR estimates that we obtained using GARCH and under the assumption of normal distribution meet the requirements of the uc test. We start by estimating the VaR at the 95% confidence level and obtaining the time series of violations:

```
. use http://www.stata-press.com/data/feus/index, clear
. quietly summarize SP500
. scalar mean = r(mean)
. quietly arch SP500, arch(1) garch(1)
. predict variance, variance
. scalar confidencelevel = 1 - 0.95
. generate VaR = mean+variance^0.5*invnormal(confidencelevel)
. generate violations = SP500<VaR
```

The time series of violations identifies the cases when observed losses are larger than those identified by VaR.

On the basis of the time series of violations just obtained, we compute the LR_{uc} test by applying (5.10):

```
. summarize violations
```

Variable	Obs	Mean	Std. Dev.	Min	Max
violations	5,828	.0528483	.2237498	0	1

```
. scalar T1 = r(sum)
. scalar T = r(N)
. scalar empiricalratio = T1/T
. display empiricalratio
.05284832
. scalar loglEmpirical = (T-T1)*log(1-empiricalratio)+T1*log(empiricalratio)
. scalar loglTheoretical = (T-T1)*log(1-confidencelevel)+T1*log(confidencelevel)
. scalar LRUC = 2*(loglTheoretical-loglEmpirical)
. display 1-chi2(1,abs(LRUC))
.32269303
. drop VaR violations
```

In the first part of the code, we have computed the empirical ratio of violations, where T1 is the number of violations and T denotes the length of the time series. We then apply the uc test. The result indicates that we cannot reject the null hypothesis that the violations obtained in our model differ from a Bernoulli with parameter 0.05; hence, our model provides a good in-sample forecast of VaR. Note that the empirical ratio of violations π is 0.0528, which is very close to the theoretical α 0.05.

Let us now consider the 99% confidence level:

```
. scalar confidencelevel = 1 - 0.99
. generate VaR = mean+variance^0.5*invnormal(confidencelevel)
. generate violations = SP500<VaR
```

```
. summarize violations
    Variable |        Obs        Mean     Std. Dev.        Min         Max
  -----------+-----------------------------------------------------------
  violations |      5,828     .0195607      .138497          0           1
. scalar T1 = r(sum)
. scalar T = r(N)
. scalar empiricalratio = T1/T
. display empiricalratio
.01956074
. scalar loglEmpirical = (T-T1)*log(1-empiricalratio)+T1*log(empiricalratio)
. scalar loglTheoretical = (T-T1)*log(1-confidencelevel)+T1*log(confidencelevel)
. scalar LRUC = 2*(loglTheoretical-loglEmpirical)
. display 1-chi2(1,abs(LRUC))
8.788e-11
. drop variance VaR violations
```

Now the empirical ratio of violations is 0.0196, which is not so close to the theoretical 0.01. The uc test does not allow us to not reject the null hypothesis of equivalence of the empirical and theoretical ratios. We can interpret this result as the inability of the normal distribution to describe data in a satisfactory way when considering a higher confidence level, that is, when moving far away from the center of the distribution toward its tails.

We now switch to the Student's *t* distribution and verify whether this distribution allows us to model in a more satisfactory way the tails of the distribution.

```
. quietly arch SP500, arch(1) garch(1) distribution(t)
. predict variance, variance
. generate VaR = mean+variance^0.5*invt(6,confidencelevel)
. generate violations = SP500<VaR
```

We fit a GARCH(1,1) model under the assumption of a Student's t distribution (results are omitted to save space). The estimated degrees of freedom is 6, the same we use when computing the VaR percentile. We now apply the uc test:

```
. summarize violations
```

Variable	Obs	Mean	Std. Dev.	Min	Max
violations	5,828	.0054907	.0739021	0	1

```
. scalar T1 = r(sum)
. scalar T = r(N)
. scalar empiricalratio = T1/T
. display empiricalratio
.00549073
. scalar loglEmpirical = (T-T1)*log(1-empiricalratio)+T1*log(empiricalratio)
. scalar loglTheoretical = (T-T1)*log(1-confidencelevel)+T1*log(confidencelevel)
. scalar LRUC = 2*(loglTheoretical-loglEmpirical)
. display 1-chi2(1,abs(LRUC))
.00015504
. drop variance VaR violations
```

The empirical ratio of violations is equal to 0.55%, which is still quite far from the 1% required by the test. Consistently, the uc test does not allow us to not reject the null hypothesis that the time series of violations we obtain under a Student's t distribution with 6 degrees of freedom is not different from a sequence of numbers extracted from a Bernoulli with parameter 0.01.

The independence test

The intuition behind the independence test (ind) is that in a good model, a VaR violation today should be independent of whether or not VaR was violated yesterday. Therefore, in the ind test, we evaluate whether the hit sequence I_t shows any serial autocorrelation. Formally, the null hypothesis is

$$H_{0,\text{ind}}\colon \pi_{01} = \pi_{11}$$

where $\pi_{r,s}$ is the probability of an r at day $t-1$ being followed by an s at day t. For instance, π_{01} denotes the case of no violation (0) followed by a violation (1), while π_{11} denotes the case of a violation (1) followed by another violation (1).

The alternative hypothesis in the ind test is that the hit sequence I_t follows a first-order Markov sequence with switching probability matrix:

$$\Pi = \begin{bmatrix} 1 - \pi_{01} & \pi_{01} \\ 1 - \pi_{11} & \pi_{11} \end{bmatrix}$$

The test statistic is then defined as

$$\mathrm{LR}_{\mathrm{ind}} = 2\left\{\ln L(z; \widehat{\pi}_{01}, \widehat{\pi}_{11}) - \ln L(z; \widehat{\pi}_1)\right\} \sim \chi_1^2 \qquad (5.11)$$

where

$$L(z; \pi_{01}, \pi_{11}) = (1 - \pi_{01})^{T0 - T01} \pi_{01}^{T01} (1 - \pi_{11})^{T1 - T11} \pi_{11}^{T11}$$

with T_{rs} being the number of observations with an r followed by an s. In addition, $\widehat{\pi}_{01} = T01/T0$ is the percentage of cases when no violation is followed by a violation, and $\widehat{\pi}_{11} = T11/T1$ is the percentage of cases when a violation is followed by another violation.

We now apply the ind test to the time series of daily returns of the S&P 500. We start by considering the case when VaR is computed using a normal GARCH model. The first step in implementing the ind test consists of obtaining the violations, a procedure that we have illustrated just above. Once we dispose of the hit sequence, we have to compute the number of days with a violation that are preceded by days with no violation (π_{01}) and the number of days with a violation that are preceded by a violation (π_{11}).

```
. quietly arch SP500, arch(1) garch(1)
. predict variance, variance
. scalar confidencelevel = 1 - 0.95
. generate VaR = mean+variance^0.5*invnormal(confidencelevel)
. generate violations = SP500<VaR
. generate violations01 = (L.violations==0 & violations==1)
. generate violations11 = (L.violations==1 & violations==1)
. summarize violations01
```

Variable	Obs	Mean	Std. Dev.	Min	Max
violations01	5,828	.0502745	.2185296	0	1

```
. scalar T01 = r(sum)
. scalar T = r(N)
. summarize violations11
```

Variable	Obs	Mean	Std. Dev.	Min	Max
violations11	5,828	.0025738	.0506715	0	1

```
. scalar T11 = r(sum)
. scalar T1 = T01+T11
. scalar T0 = T-T1
. scalar pi01 = T01/T0
. scalar pi11 = T11/T1
. scalar pi1 = T1/T
```

At this point, we can compute the two log likelihoods and apply the ind test:

```
. scalar loglEmpirical = (T0-T01)*log(1-pi01)+T01*log(pi01)+(T1-T11)*
> log(1-pi11)+T11*log(pi11)
. scalar loglTheoretical = (T-T1)*log(1-pi1)+T1*log(pi1)
. scalar LRInd = 2*(loglTheoretical-loglEmpirical)
. display 1-chi2(1,abs(LRInd))
.73514595
. drop violations01 violations11
```

The results indicate that we cannot reject the null hypothesis that the theoretical and the empirical log likelihood differ so much. We can therefore conclude that our VaR model at the 95% confidence level is correctly specified, even as far as the independence of the VaR violations is concerned.

The conditional coverage test

Neither the uc test nor the ind test is complete on its own, given that the first one evaluates whether on average the coverage rate α of the VaR model is correct, while the second focuses mainly on the clustering effect on the failures sequence. The conditional coverage (cc) test combines both, evaluating the null hypothesis

$$H_{0,cc}: \pi_{01} = \pi_{11} = \alpha$$

The null hypothesis implies that the percentage of violations should be equal to the theoretical confidence level α, regardless of whether violations are preceded by another violation.

The likelihood-ratio test for the cc is

$$\mathrm{LR}_{cc} = 2\left\{\ln L(z; \widehat{\pi}_{01}, \widehat{\pi}_{11}) - \ln L(z; p)\right\} \sim \chi_2^2$$

We apply the cc test to the time series of the S&P 500 daily returns. In the first step, we obtain the time series of violations. Then, and following the same steps as the ind test, we compute the quantities π_{01}, π_{11}, and π_1.

```
. generate violations01 = (L.violations==0 & violations==1)
. generate violations11 = (L.violations==1 & violations==1)
. summarize violations01
```

Variable	Obs	Mean	Std. Dev.	Min	Max
violations01	5,828	.0502745	.2185296	0	1

```
. scalar T01 = r(sum)
. scalar T = r(N)
. summarize violations11
```

Variable	Obs	Mean	Std. Dev.	Min	Max
violations11	5,828	.0025738	.0506715	0	1

```
. scalar T11 = r(sum)
. scalar T1 = T01+T11
. scalar T0 = T-T1
. scalar pi01 = T01/T0
. scalar pi11 = T11/T1
. scalar pi1 = T1/T
```

Finally, we compute the two log likelihoods, the empirical and the theoretical, where the empirical log likelihood is (5.11) and the theoretical is (5.9):

```
. scalar loglEmpirical = (T0-T01)*log(1-pi01)+T01*log(pi01)+(T1-T11)*
> log(1-pi11)+T11*log(pi11)
. scalar loglTheoretical = (T-T1)*log(1-confidencelevel)+T1*log(confidencelevel)
. scalar LRCC = 2*(loglTheoretical-loglEmpirical)
. display 1-chi2(2,abs(LRCC))
.57913505
```

Even the conditional coverage test supports the finding that the model is correctly specified. This result was somehow expected because we find that both the uc and the ind test indicates that the VaR violations from our model satisfy the desired properties. Hence, given that the cc test is somewhat a summary of the two other tests, it should go in the same direction as the two individual tests.

The duration tests

Christoffersen and Pelletier (2004) propose a generalization of the ind test considering a broader alternative compared with the Markov first order. Therefore, to apply this test, we should first define the duration between two VaR violations (the no-hit duration) as

$$D_s = t(I_s) - t(I_{s-1})$$

where $t(I_s)$ denotes the time interval of violation number s.

Under the null hypothesis that the risk model is correctly specified, the no-hit duration should have no memory and a mean duration of $1/p$ time intervals; the distribution satisfying the memoryless property is the exponential distribution.

$$f_{\mathrm{exp}}(D;p) = p\exp(-pD)$$

As an alternative distribution, they select the Weibull because it allows for duration dependence as well as being a generalization of the exponential:

$$f_{\mathrm{exp}}(D;p) = p^b b D^{b-1} \exp\left\{-(pD)^b\right\}$$

The hypothesis we want to test is therefore

$$H_{0,\mathrm{DurWeibull}}: b = 1$$

The main drawback of this test is that it does not analyze the temporal ordering of a no-hit duration.

To overcome this limitation, Christoffersen and Pelletier (2004) propose another test relying on the exponential autoregressive conditional duration (EACD) model by Engle and Russell (1998). The EACD(0,1) model takes the form

$$E_{t-1}(D_i) = \psi = \omega + \gamma D_{i-1}$$

with $\gamma \in [0,1)$. Assuming an underlying exponential density with a mean of 1, the conditional distribution of the density is

$$f_{\mathrm{EACD}}(D_i; \frac{1}{\psi_i}) = \frac{1}{\psi_i}\exp\left(-\frac{D_i}{\psi_i}\right)$$

The null of independent no-hit duration would then correspond to $H_0 : \gamma = 0$.

We would like to point out that while the large-sample distribution of the likelihood-ratio tests described above is chi-squared, the dearth of violations of 1% VaR make the effective sample size rather small, even when the nominal size is large. To overcome this problem and to obtain p-values robust to finite-sample scenarios, Christoffersen and Pelletier (2004) suggest to adopt the Monte Carlo tests of Dufour (2006). This procedure consists of generating M independent realizations—1,000, for instance—for each one of the four test statistics: $\mathrm{LR}_{i,type}$, where $i = 1, \ldots, M$ and $type = \mathrm{uc, ind, cc, DurWeibull}$. The case $\mathrm{LR}_{0,type}$ corresponds to the calculated test statistic. The Monte Carlo p-value $\widehat{p}_M(\mathrm{LR}_0)$ is thus

$$\widehat{p}_M(\mathrm{LR}_0) = \frac{\widehat{G}_M(\mathrm{LR}_0) + 1}{M + 1}$$

where

$$\widehat{G}_M(\mathrm{LR}_0) = M - \sum_{i=1}^{M}\mathbf{I}(\mathrm{LR}_i < \mathrm{LR}_0) + \sum_{i=1}^{M}\mathbf{I}(\mathrm{LR}_i = \mathrm{LR}_0)\mathbf{I}(U_i \geq U_0)$$

where $\mathbf{I}(\cdot)$ is the indicator function and U_i, $i = 0, \ldots, M$, are independent realizations of a uniform distribution on the $[0, 1]$ interval.

6 Contagion analysis

6.1 Introduction

Until the 1980s, financial crises were considered events that occurred in individual markets, without a systemic nature. This explains why economic and financial literature on the transmission mechanism of shocks had little attention from researchers until the 1990s, when several financial crises such as the crisis of the European Mechanism Exchange Rates (1992), the Mexican crisis (1994–1995), the Asian crisis (1997–1998), the Russian crisis (1998), and the Brazilian crisis (1999) occurred. One of the most impressive characteristics of these crises was that the time when they took place and their intensity did not seem to be related to the fundamental problems that those countries were facing. Furthermore, the negative consequences associated with the instability episodes were not limited to the countries of origin; they were quickly transmitted all over the world to several international markets. The literature on contagion start to develop at this stage.

In what follows, we briefly revise the literature on contagion with particular focus on some econometric methodologies proposed to test for the presence of contagion among stock markets.

Let us start by defining the concept of contagion. Though an exact definition in literature does not exist, the most accepted one is derived from the seminal contribution by Forbes and Rigobon (2002), who define contagion as a significant increase in cross-markets linkage after a shock to one country or to a group of countries. What distinguishes contagion from simple interdependence among stock markets is the significant increase in cross-markets linkage: if after a shock two markets show a high degree of comovement that also characterized the analyzed markets during the period of stability preceding the crisis, then this is not contagion but a simple interdependence.

It is important to study the transmission of shocks among international financial markets for different reasons. First, contagion may have deep implications for portfolio management, particularly in the process of international diversification of risk. Second, the importance of studying contagion has been reinforced by the tendency for integrating financial markets worldwide. Third, the study of contagion is central to better understand the role and the effectiveness of the interventions of the financial institutions in contexts of crisis.

The literature on contagion is based on two main causes for contagion to take place. We define fundamental-based contagion when the country hit by a financial crisis is linked to the others via trade or finance. We refer to pure contagion when common shocks through trade or finance are either not present or have been controlled for. Policymakers need to be aware of the source of contagion when they have to put in place appropriate responses to a crisis. If the cause of a crisis is a random jump between equilibrium—that is, contagion—then international institutional lending to prevent contagion could be a highly effective response because it might return the market to the "good" equilibrium. If, in contrast, a crisis spreads to other countries because their fundamentals are correlated or there are spillovers affecting the economic fundamentals, then international institutional lending cannot prevent the crisis unless it is large enough to change the fundamentals.

Finally, when testing for the presence of contagion, it is necessary to consider that a crash of stock markets does not affect markets just in the short-run but could cause a dramatic loss of confidence for investors in stock exchanges, capital flights, and speculative runs that could lead to a more prolonged financial instability period.

In 2008–2009, the United States experienced the worst financial crisis since the Great Depression. This crisis began in the United States with the bursting of the subprime mortgage market and the unrevealing of the securitization process in the summer of 2007 that had been built up to finance those mortgages. It soon became clear that this crisis would not only have had deep consequences in the United States, but that each region of the globe would have been influenced by it. Across the world, stock prices fell, volatility levels jumped, credit spreads increased sharply, and liquidity demand surged, prompting central banks to inject substantial amounts of additional liquidity into the markets. Uncertainty was very pronounced in the short-term money markets, as evidenced by a marked increase in risk aversion in the asset-backed commercial paper market and rather unprecedented rises in interbank money market interest rates. At the same time, markets for the transfer of credit risk were affected as well, and the costs of credit protection increased markedly, particularly for specific sectors such as banks and other financial institutions.

Although individual countries and specific market segments were affected at different extents, the financial market turmoil was truly global, and discrimination by market participants among individual borrowers was uneven or nonexistent.

The financial turmoil revealed several weaknesses related to the use of structured finance that can be summarized as follows. In numerous cases, banks underestimated their exposures to structured finance products and to specific off-balance sheet vehicles that played an important role in this type of finance. Moreover, certain banks invested heavily in structured finance products, retaining large exposures to specific structured finance instruments such as collateralized debt obligations but without understanding sufficiently their impact on the banks' capital and liquidity positions. When the financial turmoil hit and structured credit markets came to a virtual standstill, the funding capability of specific banks, such as Northern Rock in the United Kingdom, was impaired significantly. Furthermore, many of the globally operating banks had offered liquidity standby facilities to off-balance sheet vehicles engaged in structured finance, but generally underestimated the liquidity risk arising from off-balance sheet exposures.

The financial turmoil has raised concerns that the process of securitization may have generated unwelcome incentive problems, in the sense that banks may not assess the credit risk of specific borrowers accurately because they put these loans off the balance sheet through securitization techniques.

Though the base of the recent crises comprised mainly financial reasons, politics also played a key role, for instance, by allowing the two government-sponsored enterprises to take on unlimited risks by lowering their mortgage standards with an implicit government guarantee. The Community Reinvestment Act of the mid-1990s establishes that banks have to use flexible or innovative methods in lending banks, to offer mortgages to under-served areas. The evidence reveals that bank mortgage standards fell as a consequence of this regulatory change. Riskier mortgage standards were not the consequence of deregulation; rather, the banks were compelled to change the standards by new regulations at the behest of community groups. Moreover, rating agencies, which rated the products that had the subprime mortgages, were paid by the same issuers of these products, creating a moral hazard problem.

With the purpose of analyzing the severity of the subprime crisis, throughout this chapter we evaluate the presence of contagion from the United States to three countries, representing three areas of the world, namely, the United Kingdom, Germany, and Japan.

6.2   Contagion measurement

There is a large body of empirical literature testing for how shocks propagate from one country to another and testing for the presence of contagion. Here we are going to focus our attention on three alternative methodologies to measure how shocks transmit internationally:

1. cross-market correlation coefficients;
2. ARCH and GARCH models; and
3. higher moments contagion.

6.2.1   Cross-market correlation coefficients

The first and easiest approach to test for contagion consists of comparing correlation coefficients of asset returns before and after the shock and looking for an eventual significant increase of them. For instance, following Forbes and Rigobon (2002), we first have to fit a VAR model to control for any idiosyncratic disturbances and fundamental links between the stock markets,

$$X_t = A(L)X_t + \varepsilon_t$$

where $X_t = (r_{t,\text{crisis}}, r_t)$ is the vector of stock returns in the crisis country and in the noncrisis country at time t, and $A(L)X_t$ is the lag vector of stock returns.

Working on VAR residuals ε_t, we are preventing estimates to be affected by some idiosyncratic disturbances. Forbes and Rigobon (2002) demonstrate that the correlation coefficient is biased because of heteroskedasticity: during the turmoil period, stock markets are more volatile and correlation coefficient estimates tend to increase and be biased upward. What they suggest is a correction of the classic Pearson correlation coefficient to be able to account for heteroskedasticity:

$$\rho = \frac{\rho^*}{\sqrt{1 + \left(\frac{\sigma_h^2}{\sigma_l^2} - 1\right)\left\{1 - (\rho^*)^2\right\}}}$$

where ρ^* is the conditional correlation coefficient, σ_h is the volatility during the turmoil period in the country where the crisis originated, and σ_l is the volatility during the relative market stability period in the country where the crisis originated.

The null hypothesis to test is that correlation coefficients computed over the two periods are similar (that is, no contagion took place), against the alternative that the correlation during the crisis is higher than in the period preceding the crisis (that is, contagion effectively took place). Therefore, we construct a t test of the form

$$\left\{\frac{\rho_h - \rho_l^*}{\sqrt{\text{var}(\rho_h - \rho_l^*)}}\right\}^2 \tag{6.1}$$

where ρ_h denotes Forbes and Rigobon's correlation during the crisis period, and ρ_l^* is Pearson's correlation during the relative stable period. This correlation test for contagion is asymptotically distributed as chi-squared with 1 degree of freedom. Here the test statistic is two-sided, which generalizes the original test proposed by Forbes and Rigobon (2002) by allowing for dependence in each period $\rho_l^* = \rho_h^* \neq 0$ and an increase in volatility in the source country $(\delta > 0)$.

Fry, Martin, and Tang (2010) derive the analytic expression of the standard error in (6.1). Consider the following:

$$\text{var}(\rho_h - \rho_l^*) = \text{var}\,(\rho_h) + \text{var}\,(\rho_l^*) - 2\text{cov}\,(\rho_h, \rho_l^*)$$

The first term takes the form

$$\text{var}\,(\rho_h) = \left[\frac{1}{2}\frac{(1+\delta)^2}{\left\{1 + \delta\left(1 - \rho_h^{*2}\right)\right\}^3}\left\{\frac{\left(2 - \rho_h^{*2}\right)\left(1 - \rho_h^{*2}\right)^2}{T_h} + \frac{\rho_h^{*2}\left(1 - \rho_h^{*2}\right)^2}{T_l}\right\}\right]$$

where δ is equal to $(\sigma_h/\sigma_l - 1)$, and T_h and T_l are the lengths of the precrisis and crisis periods.

The second term is

$$\text{var}\,(\rho_l^*) = \frac{1}{T_l}\left(1 - \rho_l^{*2}\right)^2$$

Finally, the third term is

$$\text{cov}\left(\rho_h, \rho_l^*\right) = \frac{1}{2}\frac{1}{T_l}\frac{\rho_l^* \rho_h^* \left(1 - \rho_h^{*\,2}\right)\left(1 - \rho_l^{*\,2}\right)(1 + \delta)}{\sqrt{\left\{1 + \delta\left(1 - \rho_h^{*\,2}\right)\right\}^3}} \qquad (6.2)$$

Forbes and Rigobon (2002) show that by adopting Pearson correlation coefficients, they find evidence of contagion in all the crises analyzed (the 1997 East Asian crisis, the 1994 Mexican peso devaluation, and the 1987 U.S. market decline) and for each country considered. However, when adjusting correlation coefficients to heteroskedasticity, the evidence of contagion disappears. Nevertheless, the authors state that even if their estimator solves the heteroskedasticity problem, it is biased in the presence of endogeneity or omitted variables.

In addition, another limitation of Forbes and Rigobon's (2002) methodology is that it requires specification of the country in which contagion originated and when the turmoil period started to be able to subdivide the preshock periods from the postshock periods.

These choices involve a certain level of discretion and can invalidate results. For example, when analyzing the East Asian crisis, it is possible to alternatively choose either Thailand or Hong Kong as the country where the crisis originated. Moreover, a shock manifested in one country can increase the correlation between two other countries, and this phenomenon cannot be evaluated by applying Forbes and Rigobon's methodology. Finally, it is important to remember that correlation coefficients allow us to measure only linear dependence between two variables.

Empirical exercise

We now evaluate the presence of contagion from the United States to Germany, the United Kingdom, and Japan corresponding to the burst of the financial crisis.

We compare the correlation coefficient computed over the precrisis period with that from the postcrisis period, setting the switching date regime on 15 September 2008, the date of the collapse of Lehman Brothers. To show the differences between the standard Pearson correlation coefficient and the Forbes and Rigobon heteroskedasticity robust version, we compute both of them.

We start by fitting a VAR(1) model with the **var** command. We then obtain the residuals with the **predict** command combined with the **residuals** and **equation()** options. Note that because the VAR model is a type of ordinary least-squares regression, the first-stage estimation error might not matter.

```
. use http://www.stata-press.com/data/feus/index
. tsset newdate
        time variable:  newdate, 02jan1991 to 30dec2013
              delta:  1 day
```

```
. var SP500 FTSE100 Dax30 Nikkei225 if date >= mdy(01,01,2002) &
> date <= mdy(12,31,2009), lags(1)
```

Vector autoregression

Sample: 02jan2002 - 30dec2009				Number of obs	=	2,038
Log likelihood =	25052.84			AIC	=	-24.56609
FPE =	2.52e-16			HQIC	=	-24.54586
Det(Sigma_ml) =	2.47e-16			SBIC	=	-24.51094

Equation	Parms	RMSE	R-sq	chi2	P>chi2
SP500	5	.01384	0.0194	40.27275	0.0000
FTSE100	5	.012438	0.1588	384.7054	0.0000
Dax30	5	.016164	0.0738	162.4207	0.0000
Nikkei225	5	.01335	0.3031	886.4721	0.0000

	Coef.	Std. Err.	z	P>|z|	[95% Conf. Interval]	
SP500						
SP500						
L1.	-.1381318	.0278413	-4.96	0.000	-.1926996	-.0835639
FTSE100						
L1.	-.0825143	.0392366	-2.10	0.035	-.1594166	-.0056119
Dax30						
L1.	.100547	.0336384	2.99	0.003	.034617	.166477
Nikkei225						
L1.	-.0476476	.0204039	-2.34	0.020	-.0876385	-.0076567
_cons	-.0000523	.0003062	-0.17	0.865	-.0006524	.0005479
FTSE100						
SP500						
L1.	.4516721	.0250205	18.05	0.000	.4026327	.5007114
FTSE100						
L1.	-.2880994	.0352614	-8.17	0.000	-.3572104	-.2189884
Dax30						
L1.	-.0326969	.0302303	-1.08	0.279	-.0919473	.0265535
Nikkei225						
L1.	-.0081712	.0183367	-0.45	0.656	-.0441104	.0277681
_cons	.0000445	.0002752	0.16	0.871	-.0004948	.0005839
Dax30						
SP500						
L1.	.3920662	.0325169	12.06	0.000	.3283341	.4557982
FTSE100						
L1.	-.1759635	.045826	-3.84	0.000	-.2657808	-.0861461
Dax30						
L1.	-.132748	.0392876	-3.38	0.001	-.2097503	-.0557456

Nikkei225						
L1.	-.0010577	.0238305	-0.04	0.965	-.0477647	.0456493
_cons	.0000998	.0003576	0.28	0.780	-.0006011	.0008008
Nikkei225						
SP500						
L1.	.4663157	.026856	17.36	0.000	.413679	.5189524
FTSE100						
L1.	.198091	.037848	5.23	0.000	.1239102	.2722717
Dax30						
L1.	.0592873	.0324479	1.83	0.068	-.0043094	.1228841
Nikkei225						
L1.	-.1270892	.0196818	-6.46	0.000	-.1656649	-.0885136
_cons	-.0000627	.0002954	-0.21	0.832	-.0006417	.0005162

```
. predict res_SP500, residuals equation(SP500)
(1 missing value generated)
. predict res_FTSE100, residuals equation(FTSE100)
(1 missing value generated)
. predict res_Dax30, residuals equation(Dax30)
(1 missing value generated)
. predict res_Nikkei225, residuals equation(Nikkei225)
(1 missing value generated)
```

We now can compute Pearson's and Forbes and Rigobon's correlation coefficients on residuals. We start by considering the pair S&P 500 and FTSE 100:

```
. summarize res_SP500 if date >= mdy(01,01,2002) & date <= mdy(09,14,2008)
```

Variable	Obs	Mean	Std. Dev.	Min	Max
res_SP500	1,707	.0000745	.0104109	-.0409247	.05624

```
. scalar sdSP500_Precrisis = r(sd)
. scalar N_precrisis = r(N)
. summarize res_SP500 if date >= mdy(09,15,2008) & date <= mdy(12,31,2009)
```

Variable	Obs	Mean	Std. Dev.	Min	Max
res_SP500	331	-.0003843	.0248909	-.0920638	.1028357

```
. scalar sdSP500_crisis = r(sd)
. scalar N_crisis = r(N)
. * Pearson correlation coefficient, precrisis
. correlate res_SP500 res_FTSE100 if date >= mdy(01,01,2002) &
> date <= mdy(09,14,2008)
(obs=1,707)
```

	res_~500 res_~100
res_SP500	1.0000
res_FTSE100	0.5233 1.0000

```
. scalar corr_precrisis = r(rho)
```

```
. * Pearson correlation coefficient, crisis
. correlate res_SP500 res_FTSE100 if date >= mdy(09,15,2008) &
> date <= mdy(12,31,2009)
(obs=331)
                 | res_~500 res_~100

   res_SP500 |    1.0000
  res_FTSE100 |   0.7090    1.0000

. scalar corr_crisis = r(rho)
. * Forbes and Rigobon correlation coefficient, crisis
. scalar delta = sdSP500_crisis^2/sdSP500_Precrisis^2-1
. scalar corrFR_crisis = corr_crisis/sqrt(1+delta*(1-corr_crisis^2))
. display "correlation precrisis = " corr_precrisis
correlation precrisis = .5232799
. display "correlation crisis = " corr_crisis
correlation crisis = .70896108
. display "correlation crisis Forbes and Rigobon = " corrFR_crisis
correlation crisis Forbes and Rigobon = .38759428
```

In the first part of the code, we compute the standard deviation of the S&P 500 daily returns for the precrisis period, defined between 1 January 2002 and 14 September 2008, and for the crisis period, defined between 15 September 2008 and 31 December 2009. In the second part of the code, we compute the correlation during the precrisis and the crisis period. In the crisis period, the standard Pearson correlation is higher than the correlation proposed by Forbes and Rigobon, 0.71 against 0.39. Considering the Pearson correlations, the correlation coefficient during the crisis period is higher compared with the precrisis period, 0.71 against 0.52; the same does not hold true when we consider the Forbes and Rigobon crisis period, with 0.39 being smaller than 0.52.

Forbes and Rigobon propose to correct the correlation coefficient only during the crisis period; therefore, we compare it against the precrisis period Pearson correlation coefficient.

Now we can compare the correlation coefficients computed over the two periods adopting a simple t test, where standard error is computed via (6.2):

```
. scalar var1 = (0.5*(1+delta)^2/(1+delta*(1-corr_crisis^2)))^3*((2-corr_crisis^2)*
> (1-corr_crisis^2)^2/N_crisis+corr_crisis^2*(1-corr_crisis^2)^2/N_precrisis))
. scalar var2 = 1/N_precrisis*(1-corr_precrisis^2)^2
. scalar cov = 0.5*1/N_precrisis*corr_crisis*corr_precrisis*(1-corr_crisis^2)*
> (1-corr_precrisis^2)*(1+delta)/sqrt((1+ delta*(1-corr_crisis^2))^3)
. scalar varTot = var1+var2-2*cov
. display "t-test H0: correlation precrisis = FR correlation crisis  = "
> ((corrFR_crisis-corr_precrisis)/varTot^0.5)^2
t-test H0: correlation precrisis = FR correlation crisis  = 24.366585
```

Because the test is two-sided and the region to reject is identified by the chi-squared distribution with 1 degree of freedom, the acceptance region of the null hypothesis at the 95% confidence level is between 0.001 and 5.024. Therefore, if the test statistic takes any value lying between these two extremes, then we do not reject the null hypothesis of no contagion; if the test statistic is outside the acceptance region, then we conclude with evidence of contagion. In our case, the test statistic is by far larger than the critical value, probably influenced by the larger annualized volatility during the crisis period compared with the normal period (40.14% versus 16.79%). Instead, we showed before that once the correlation coefficient during the crisis is adjusted by volatility as requested by Forbes and Rigobon (2002), it is lower than the correlation coefficient during the calm period.

We now evaluate the presence of contagion from the United States to Germany and Japan.

```
. ****************** DAX ******************************************************
. * Pearson correlation coefficient, precrisis
. correlate res_SP500 res_Dax30 if date >= mdy(01,01,2002) & date <=
> mdy(09,14,2008)
(obs=1,707)

             | res_~500 res_D~30
-------------+------------------
   res_SP500 |  1.0000
   res_Dax30 |  0.6197   1.0000

. scalar corr_precrisis = r(rho)

. * Pearson correlation coefficient, crisis
. correlate res_SP500 res_Dax30 if date >= mdy(09,15,2008) & date <=
> mdy(12,31,2009)
(obs=331)

             | res_~500 res_D~30
-------------+------------------
   res_SP500 |  1.0000
   res_Dax30 |  0.7608   1.0000

. scalar corr_crisis = r(rho)

. * Forbes and Rigobon correlation coefficient, crisis
. scalar corrFR_crisis = corr_crisis/sqrt(1+delta*(1-corr_crisis^2))

. display "correlation precrisis = " corr_precrisis
correlation precrisis = .61974145

. display "correlation crisis = " corr_crisis
correlation crisis = .76084487

. display "correlation crisis Forbes and Rigobon= " corrFR_crisis
correlation crisis Forbes and Rigobon= .44029745

. scalar var1 = (0.5*(1+delta)^2/(1+delta*(1-corr_crisis^2))^3*((2-corr_crisis^2)*
> (1-corr_crisis^2)^2/N_crisis+corr_crisis^2*(1-corr_crisis^2)^2/N_precrisis))

. scalar var2 = 1/N_precrisis*(1-corr_precrisis^2)^2

. scalar cov = 0.5*1/N_precrisis*corr_crisis*corr_precrisis*(1-corr_crisis^2)*
> (1-corr_precrisis^2)*(1+delta)/sqrt((1+ delta*(1-corr_crisis^2))^3)

. scalar varTot = var1+var2-2*cov

. display "t-test H0: correlation precrisis = FR correlation crisis  = "
> ((corrFR_crisis-corr_precrisis)/varTot^0.5)^2
t-test H0: correlation precrisis = FR correlation crisis  = 49.771859
```

```
. ******************* NIKKEI *****************************************************
. * Pearson correlation coefficient, precrisis
. correlate res_SP500 res_Nikkei225 if date >= mdy(01,01,2002) & date <=
> mdy(09,14,2008)
(obs=1,707)

             | res_~500 res_~225

    res_SP500 |    1.0000
 res_Nikk~225 |    0.1568    1.0000

. scalar corr_precrisis = r(rho)

. * Pearson correlation coefficient, crisis
. correlate res_SP500 res_Nikkei225 if date >= mdy(09,15,2008) & date <=
> mdy(12,31,2009)
(obs=331)

             | res_~500 res_~225

    res_SP500 |    1.0000
 res_Nikk~225 |    0.2649    1.0000

. scalar corr_crisis = r(rho)

. * Forbes and Rigobon correlation coefficient, crisis
. scalar corrFR_crisis = corr_crisis/sqrt(1+delta*(1-corr_crisis^2))

. display "correlation precrisis = " corr_precrisis
correlation precrisis = .15682742

. display "correlation crisis = " corr_crisis
correlation crisis = .26489024

. display "correlation crisis Forbes and Rigobon= " corrFR_crisis
correlation crisis Forbes and Rigobon= .11414615

. scalar var1 = (0.5*(1+delta)^2/(1+delta*(1-corr_crisis^2))^3*
> ((2-corr_crisis^2)* (1-corr_crisis^2)^2/N_crisis+corr_crisis^2*
> (1-corr_crisis^2)^2/N_precrisis))

. scalar var2 = 1/N_precrisis*(1-corr_precrisis^2)^2

. scalar cov = 0.5*1/N_precrisis*corr_crisis*corr_precrisis*(1-corr_crisis^2)*
> (1-corr_precrisis^2)*(1+delta)/sqrt((1+ delta*(1-corr_crisis^2))^3)

. scalar varTot = var1+var2-2*cov

. display "t-test H0: correlation precrisis = FR correlation crisis  = "
> ((corrFR_crisis-corr_precrisis)/varTot^0.5)^2
t-test H0: correlation precrisis = FR correlation crisis  = 1.6894305
```

Focusing our attention on Germany, we find a similar result as for the United Kingdom: the test statistic lies outside the acceptance region, leading us to reject the null hypothesis about the absence of contagion mainly because of the higher volatility of the S&P 500 during the turmoil period versus the calm one. In the case of Japan, the test statistic is equal to 1.69, lying in the acceptance region and implying that we do not reject the null hypothesis about absence of contagion from the United States.

6.2.2 ARCH and GARCH models

A second approach to estimate the variance–covariance transmission mechanism between countries is to use the ARCH and GARCH framework. An example in literature is the approach proposed by Chiang, Jeon, and Li (2007), who suggest using a DCC model to test for evidence of contagion during the Asian crisis of 1997. This methodology consists of estimating the time-varying correlation matrix of standardized residuals of the VAR model concerning the level of asset returns. In this way, it is possible to face the problem of heteroskedasticity raised by Forbes and Rigobon (2002) without arbitrarily dividing the sample into two subperiods. This is an important advantage over Forbes and Rigobon's estimator. Moreover, this technique does not require specification of the country where the crisis originated because all coefficients of the correlation matrix are estimated. This makes it possible to test for an increase in correlations among each pair of countries and not just between the country where the crisis started and the other countries, as is the case with Forbes and Rigobon's approach.

The Chiang, Jeon, and Li (2007) model is based on a system of equations describing the trajectories of asset returns:

$$R_t = \gamma_0 + \gamma_1 R_{t-1} + \gamma_2 R_{t-1}^O + \varepsilon_t$$

where R_t is the vector of stock market returns where the crisis could have been transmitted at time t, and R_{t-1}^O is the vector of returns of the country where the crisis originated. Finally, the vector of residuals at time t is conditionally distributed as a normal variable with a mean vector of 0 and a variance–covariance matrix H_t. The variance–covariance matrix H_t is specified as per the DCC model:

$$H_t = D_t R_t D_t$$

Once we have estimated correlations, we need to develop a test for contagion. Chiang, Jeon, and Li (2007) propose using three dummy variables for different subsamples, which is useful to investigate the dynamic feature of the correlation changes associated with different phases of the crisis. The regression model is

$$\rho_{ij,t} = \sum_{p=1}^{P} \phi_p \rho_{ij,t-p} + \sum_{k=1}^{3} \alpha_k \mathrm{DM}_{k,t} + \varepsilon_{ij,t}$$

where $\rho_{ij,t}$ is the correlation coefficient between the ith country (where the crisis originated) and the jth country (where we could find evidence of contagion), and P is the lag length determined with the AIC, giving the possibility to simultaneously take into account variance explained by regression and the number of lags included. Finally, $\mathrm{DM}_{k,t}$ with $k = 1, 2, 3$ are dummy variables identifying the precrisis, the crisis, and the postcrisis periods, respectively. Beyond the analysis of the level of asset returns, Chiang, Jeon, and Li (2007) also examine the behavior of volatility equations to detect evidence of contagion. The underlying assumption is that volatility follows a GARCH(1,1) model including three dummy variables $\mathrm{DM}_{k,t}$ with $k = 1, 2, 3$, as previously:

$$h_{ij,t} = A_0 + A_1 h_{ij,t-1} + B_1 \varepsilon_{ij,t-1}^2 + \sum_{k}^{3} \mathrm{DM}_{k,t}$$

Two important points are immediately apparent by comparing this approach with the Forbes and Rigobon approach. First, their constant correlation model fails to reveal the time-varying nature of correlation and, hence, is unable to reflect the dynamic market conditions. Second, the estimated coefficient for the constant correlation model, even with an adjustment for heteroskedasticity, is conditional on the sample size of the regime or the length of the window. The Chiang, Jeon, and Li (2007) results are interesting because they undermine portfolio diversification, one of the most classic results of finance theory: if markets are so correlated, benefits from diversification at an international level become less important because markets are subject to the same sources of systematic risks.

Empirical exercise

We start the empirical exercise by fitting a DCCT model for the S&P 500, FTSE 100, DAX 30, and Nikkei 225 for the years 2002–2009. We adopt the Tse and Tsui version because as we saw in chapter 4, it provided the best fit for this dataset.

```
. mgarch vcc Dax30 FTSE100 Nikkei225 SP500 if date >= mdy(01,01,2002) &
> date <= mdy(12,31,2009), arch(1) garch(1) nolog
Varying conditional correlation MGARCH model
Sample: 02jan2002 - 30dec2009               Number of obs   =       2,038
Distribution: Gaussian                      Wald chi2(.)    =           .
Log likelihood =  26437.18                  Prob > chi2     =           .
```

	Coef.	Std. Err.	z	P>|z|	[95% Conf.	Interval]
Dax30						
_cons	.0010873	.0002189	4.97	0.000	.0006583	.0015163
ARCH_Dax30						
arch						
L1.	.0760666	.0075782	10.04	0.000	.0612135	.0909197
garch						
L1.	.9156193	.0078192	117.10	0.000	.900294	.9309446
_cons	1.95e-06	4.00e-07	4.88	0.000	1.17e-06	2.73e-06
FTSE100						
_cons	.0007206	.0001681	4.29	0.000	.0003911	.00105
ARCH_FTSE100						
arch						
L1.	.0822925	.008235	9.99	0.000	.0661521	.0984328
garch						
L1.	.9111876	.0082348	110.65	0.000	.8950476	.9273276
_cons	1.12e-06	2.47e-07	4.52	0.000	6.32e-07	1.60e-06
Nikkei225						
_cons	.0006486	.0002537	2.56	0.011	.0001513	.0011458

ARCH_Nikk~225						
arch						
L1.	.0870619	.0112317	7.75	0.000	.0650482	.1090756
garch						
L1.	.9067751	.0112706	80.46	0.000	.8846851	.928865
_cons	2.17e-06	7.16e-07	3.04	0.002	7.71e-07	3.58e-06
SP500						
_cons	.0005822	.0001788	3.26	0.001	.0002318	.0009325
ARCH_SP500						
arch						
L1.	.0668005	.0081091	8.24	0.000	.050907	.0826939
garch						
L1.	.9231933	.0087615	105.37	0.000	.9060212	.9403655
_cons	1.17e-06	2.70e-07	4.33	0.000	6.40e-07	1.70e-06
corr(Dax30,						
FTSE100)	.8793062	.0141044	62.34	0.000	.8516621	.9069504
corr(Dax30,						
Nikkei225)	.4012136	.0454368	8.83	0.000	.3121591	.4902682
corr(Dax30,						
SP500)	.6560845	.0322829	20.32	0.000	.5928112	.7193577
corr(FTSE100,						
Nikkei225)	.3971274	.0445958	8.91	0.000	.3097212	.4845336
corr(FTSE100,						
SP500)	.5910595	.0359599	16.44	0.000	.5205793	.6615397
corr(Nik~225,						
SP500)	.2106655	.0477343	4.41	0.000	.1171081	.304223
Adjustment						
lambda1	.0137536	.0023525	5.85	0.000	.0091429	.0183643
lambda2	.9731412	.0054738	177.78	0.000	.9624128	.9838697

```
. predict corr*, correlation
```

After fitting the DCCT model, we obtain correlations by specifying the `correlation` option with the `predict` command. We type `corr*` just after `predict` to obtain correlations for all the pairs of the countries included in the sample.

Now we represent the dynamic correlations obtained by fitting the DCCT model, highlighting the Lehman Brothers collapse (see figure 6.1).

```
. tsset date
        time variable:  date, 1/2/1991 to 12/30/2013, but with gaps
                delta:  1 day
. tsline corr_SP500_FTSE100 if date >= mdy(01,01,2002) & date <= mdy(12,31,2009),
> tline(15sep2008) name(SPFTSE)
. tsline corr_SP500_Dax30 if date >= mdy(01,01,2002) & date <= mdy(12,31,2009),
> tline(15sep2008) name(SPDax)
. tsline corr_SP500_Nikkei225 if date >= mdy(01,01,2002) & date <= mdy(12,31,2009),
> tline(15sep2008) name(SPNikkei)
. graph combine SPFTSE SPDax SPNikkei
```

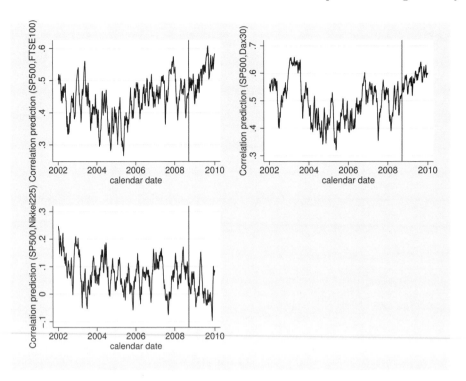

Figure 6.1. DCCT correlations between the United States, the United Kingdom, Germany, and Japan for the years 2002–2009; the vertical line identifies the Lehman Brothers collapse of 15 September 2008

With the purpose of highlighting the Lehman Brothers collapse, we specify the `tline(.)` option taking as its argument 15 September 2008. In addition, to put all the charts together, we first save the individual charts by using the `name()` option, and then we use `graph combine` to recall all the individual charts.

Figure 6.1 shows the strong increase in correlations between the S&P 500, FTSE 100, and DAX 30 experienced in the aftermath of the Lehman Brothers collapse. The same does not hold true for Nikkei 225, where we note no change in the time pattern of dynamic correlations.

To properly compare the correlation distributions, before and after the Lehman Brothers collapse, we adopt the Kolmogorov–Smirnov test by using the `ksmirnov` command.

```
. generate crisis=1 if date >= mdy(01,01,2002) & date <= mdy(09,14,2008)
(4,121 missing values generated)
. replace crisis=2 if date >= mdy(09,15,2008) & date <= mdy(12,31,2009)
(331 real changes made)
. ksmirnov corr_SP500_FTSE100 if date >= mdy(01,01,2002) & date <=
> mdy(12,31,2009), by(crisis)
Two-sample Kolmogorov-Smirnov test for equality of distribution functions
   Smaller group       D        P-value

   1:                0.6328     0.000
   2:                0.0000     1.000
   Combined K-S:     0.6328     0.000
. ksmirnov corr_SP500_Dax30 if date >= mdy(01,01,2002) & date <=
> mdy(12,31,2009), by(crisis)
Two-sample Kolmogorov-Smirnov test for equality of distribution functions
   Smaller group       D        P-value

   1:                0.6608     0.000
   2:               -0.0432     0.355
   Combined K-S:     0.6608     0.000
Note: Ties exist in combined dataset;
      there are 2037 unique values out of 2038 observations.
. ksmirnov corr_SP500_Nikkei225 if date >= mdy(01,01,2002) & date <=
> mdy(12,31,2009), by(crisis)
Two-sample Kolmogorov-Smirnov test for equality of distribution functions
   Smaller group       D        P-value

   1:                0.0000     1.000
   2:               -0.4446     0.000
   Combined K-S:     0.4446     0.000
```

In the first part of the code, we create a dummy variable, `crisis`, taking a value
of 1 for the precrisis period (1 January 2002–14 September 2008) and a value of 2 for
the crisis period (15 September 2008–31 December 2009).

In the second part, we test for differences in the distribution. `ksmirnov` proposes
three tests. In the first line, the null hypothesis is that the distribution of the first group
is stochastically dominated by the distribution of the second group, while the second
line tests whether the distribution of the second group is stochastically dominated by
the distribution of the first group. Finally, the `Combined K-S` line evaluates the equality
of the two distributions. We find strong evidence that the distribution of the correlation
coefficients during the crisis period stochastically dominates the correlations during the
precrisis period between the United States and the United Kingdom as well as the
United States and Germany. In the correlations between the United States and Japan,
however, the test suggests that the distribution of the correlation coefficient during the
precrisis period dominates the one during the crisis, not supporting the presence of
contagion.

To compare correlations graphically, we obtain the kernel distribution of the pairwise
correlations during the two subsamples, before and during the financial crisis. We then
plot the two distributions to appreciate any differences in the two subsamples. We use

the `kdensity` command to obtain the kernel distribution of the correlation between the S&P 500 and FTSE 100 in a first subsample, 1 January 2002–14 September 2008, which we store in the `kernel_pre_crisis` variable specified in the `generate()` option. We do the same for the period 15 September 2008–31 December 2009, which we store in the `kernel_crisis` variable. Having generated the two distributions, we now compare them in a single chart using the `line` command. We conduct a similar analysis for correlation between the United States and Germany as well as the United States and Japan. We report the full comparison in figure 6.2.

```
. * SP500 FTSE
. kdensity corr_SP500_FTSE100 if date >= mdy(01,01,2002) &
> date <= mdy(09,14,2008), nograph generate(kernel_pre_crisis) at (x)

. kdensity corr_SP500_FTSE100 if date >= mdy(09,15,2008) &
> date <= mdy(12,31,2009), nograph generate(kernel_crisis) at (x)

. label var kernel_pre_crisis "S&P 500, FTSE 100 Precrisis"

. label var kernel_crisis "S&P 500, FTSE 100 Crisis"

. line kernel_pre_crisis kernel_crisis x, legend(col(1)) name(KSSPFTSE)

. drop kernel_pre_crisis kernel_crisis x

. kdensity corr_SP500_Dax30 if date >= mdy(01,01,2002) & date <= mdy(09,14,2008),
> nograph generate(x kernel_pre_crisis)

. kdensity corr_SP500_Dax30 if date >= mdy(09,15,2008) & date <= mdy(12,31,2009),
> nograph generate(kernel_crisis) at (x)

. label var kernel_pre_crisis "S&P 500, Dax 30 Precrisis"

. label var kernel_crisis "S&P 500, Dax 30 Crisis"

. line kernel_pre_crisis kernel_crisis x, legend(col(1)) name(KSSPDAX)

. * SP500 NIKKEI
. drop kernel_pre_crisis kernel_crisis x

. kdensity corr_SP500_Nikkei225 if date >= mdy(01,01,2002) &
> date <= mdy(09,14,2008), nograph generate(x kernel_pre_crisis)

. kdensity corr_SP500_Nikkei225 if date >= mdy(09,15,2008) &
> date <= mdy(12,31,2009), nograph generate(kernel_crisis) at (x)

. label var kernel_pre_crisis "S&P 500, Nikkei 225 Precrisis"

. label var kernel_crisis "S&P 500, Nikkei 225 Crisis"

. line kernel_pre_crisis kernel_crisis x, legend(col(1)) name(KSSPNikkei)

. drop kernel_pre_crisis kernel_crisis x

. graph combine KSSPFTSE KSSPDAX KSSPNikkei
```

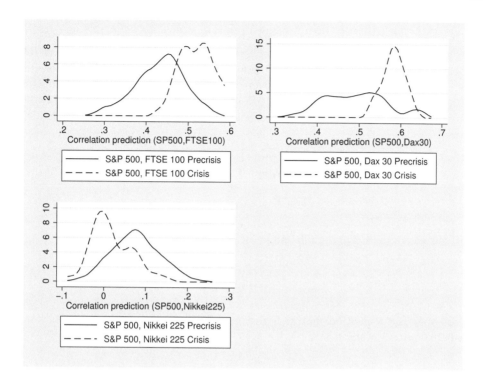

Figure 6.2. Correlation kernel distribution in the two subperiods, before and after the Lehman Brothers collapse of 15 September 2008

Figure 6.2 illustrates the different distributions characterizing the tranquil period and the turmoil one. We can see that, for the United Kingdom and Germany, the crisis distribution is much more right-tailed than the precrisis one, suggesting a general higher correlation with the United States than in normal periods. In contrast, Japan shows a more left-tailed distribution during the crisis period, evidencing a decreasing correlation of Nikkei 225 with the S&P 500 during that period.

Markov switching

An additional tool that we can adopt to analyze the time dynamics of correlations fit by using a DCC model are regime-switching models. Regime-switching models were developed to capture the tendency of many economic and financial variables to behave quite differently during the economic cycle. For instance, a variable y_t can be characterized by a state-specific time evolution

$$y_t = \mu_1 + \phi y_{t-1} + \varepsilon_t$$
$$y_t = \mu_2 + \phi y_{t-1} + \varepsilon_t$$

where μ_1 and μ_2 are the constant terms in the two possible states of the world.

The key point is that we do not know a priori in which state variable y_t is at time t.

A suitable assumption about the unobserved state of the process is that it follows a Markov chain. Therefore, regime-switching models are known as Markov-switching models. The Markov-switching regression model was initially developed in Quandt (1972) and Goldfeld and Quandt (1973). In a seminal paper, Hamilton (1989) extended Markov-switching regressions for AR processes and provided a nonlinear filter for estimation.

According to a Markov chain, the probability of transition from state i to state j in one time period is given by the following:

$$P\left(s_t = j, s_{t-1} = i\right) = p_{ij}$$

All the so-computed probabilities can be collected in the transition matrix:

$$P = \begin{bmatrix} p_{ii} & p_{ij} \\ p_{ji} & p_{jj} \end{bmatrix}$$

Note that the transition matrix is designed in such a way that for each row, the following condition holds true:

$$\sum_j p_{ij} = 1$$

In Stata, two alternative specifications of Markov-switching models are available: Markov-switching dynamic regression (MSDR) models, allowing a quick adjustment after the process changes state, and Markov-switching autoregression (MSAR) models, allowing a more gradual adjustment.

The MSDR can be represented as

$$y_t = \mu_s + x_t \alpha + z_t \beta_s + \varepsilon_s$$

where y_t is the dependent variable, μ_s is the state-dependent intercept, x_t is a vector of exogenous variables with state-invariant coefficients α, z_t is a vector of exogenous variables with state-dependent coefficients β_s, and ε_s is an i.i.d. normal error with a mean of 0 and a state-dependent variance σ_s^2. x_t and z_t may contain lags of y_t.

To illustrate the Markov-switching regression, we now focus our attention on the correlation between the S&P 500 and DAX 30 during the years 2002–2009 that we obtained in the DCCT model (see figure 6.3):

```
. tsline corr_SP500_Dax30 if date >= mdy(01,01,2002) & date <= mdy(12,31,2009)
```

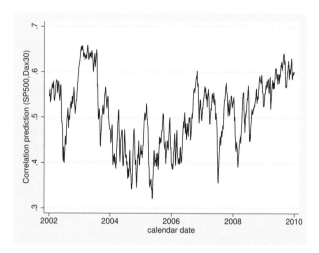

Figure 6.3. Correlation between S&P 500 and DAX 30 for the years 2002–2009

We start by fitting a two-state model on the S&P 500 and DAX 30 correlation, where the constant term is state dependent:

$$\rho_t = \mu_s + \varepsilon_t \tag{6.3}$$

```
. tsset newdate
        time variable:  newdate, 02jan1991 to 30dec2013
                delta:  1 day
. mswitch dr corr_SP500_Dax30 if date >= mdy(01,01,2002) & date <= mdy(12,31,2009)

Performing EM optimization:

Performing gradient-based optimization:

Iteration 0:    log likelihood =    3539.583
Iteration 1:    log likelihood =    3539.6026
Iteration 2:    log likelihood =    3539.6056
Iteration 3:    log likelihood =    3539.6056

Markov-switching dynamic regression

Sample: 02jan2002 - 30dec2009                No. of obs      =        2,038
Number of states =    2                      AIC             =      -3.4687
Unconditional probabilities: transition      HQIC            =      -3.4636
                                             SBIC            =      -3.4549

Log likelihood = 3539.6056
```

corr_SP50~30	Coef.	Std. Err.	z	P>|z|	[95% Conf. Interval]	
State1						
_cons	.4348885	.0017061	254.90	0.000	.4315445	.4382324
State2						
_cons	.5628561	.0014498	388.24	0.000	.5600146	.5656976
sigma	.0413063	.0006518			.0400483	.0426038
p11	.9901669	.0033236			.98098	.9949393
p21	.0069535	.0024866			.0034452	.0139843

The header reports the sample size, fit statistics, number of states, and method used for computing the unconditional state probabilities. The EM algorithm is used to find the starting values for the quasi-Newton optimizer, and we see that it took three iterations for the model to converge.

The table shows the two state-dependent constants and the error constant variance. We can see that the correlation for state 1 is on average equal to 0.43, and the correlation for state 2 is on average equal to 0.56. Thus, we can interpret state 1 as the lower correlation state and state 2 as the higher correlation state.

The lower part of the table reports the first $(k-1)$ rows of the transition matrix, where k is the number of states. We can see that state 1 is highly persistent with p11 equal to 0.99. We read this result as the following: the probability that the correlation will stay in state 1 tomorrow given that it is in state 1 today is equal to 99%. The second row, labeled p21, indicates the probability that the correlation will be in state 1 tomorrow given that it is in state 2 today. Therefore, we can obtain the probability of staying in state 2 as $1 - 0.0070 = 0.9930$, or 99.3%.

By default, the mswitch command considers two states. We can specify our own number of states with the states() option. However, we should be aware that, especially when choosing a higher number of states, convergence is not always guaranteed because it can happen that parameters of the specified model cannot be identified by the data.

We can also display the full transition matrix by using the `estat transition` command:

```
. estat transition
Number of obs = 2,038
```

Transition Probabilities	Estimate	Std. Err.	[95% Conf. Interval]	
p11	.9901669	.0033236	.98098	.9949393
p12	.0098331	.0033236	.0050607	.01902
p21	.0069535	.0024866	.0034452	.0139843
p22	.9930465	.0024866	.9860157	.9965548

We can obtain information on the average duration of the states with the `estat duration` command:

```
. estat duration
Number of obs = 2,038
```

Expected Duration	Estimate	Std. Err.	[95% Conf. Interval]	
State1	101.6974	34.37424	52.57635	197.6011
State2	143.8129	51.42796	71.50888	290.262

The table reported above informs us that state 1 lasts on average 102 days, while state 2 lasts on average 144 days. These results are coherent with the slightly higher persistence of state 2 compared with state 1 highlighted by the transition matrix.

Finally, we can obtain the in-sample probabilities of the two states by using the `predict` command with the `pr` and `smethod(filter)` options (see figure 6.4):

```
. predict state1prob if date >= mdy(01,01,2002) & date <= mdy(12,31,2009), pr
> smethod(filter)
. generate state2prob = 1-state1prob
(3,790 missing values generated)
. label var state1prob "State 1"
. label var state2prob "State 2"
. tsset date
        time variable:  date, 1/2/1991 to 12/30/2013, but with gaps
                delta:  1 day
. tsline corr_SP500_Dax30 state1prob state2prob if date >= mdy(01,01,2002) &
> date <= mdy(12,31,2009)
```

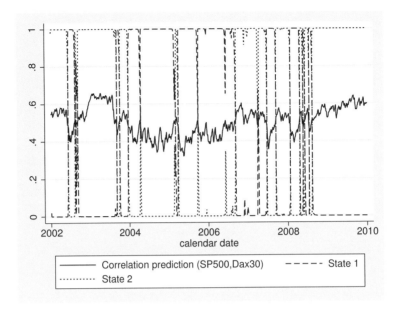

Figure 6.4. Filtered state probabilities for S&P 500 and DAX 30 correlation

From the chart reported above, we can conclude that the correlation pattern moves between two states: state 1, the long-dashed line, denoting the lower correlation periods, and state 2, the short-dashed line, denoting the higher correlation periods. We see that state 2 prevails during the two distressed periods that the stock markets experienced during the years 2002–2003, corresponding to the burst of the dot-com bubble, and the years 2008–2009, corresponding to the burst of the subprime crisis. Years 2004–2007, however, belong to state 1, characterized by a higher level of correlations and a relatively more tranquil phase experienced by the stock markets. We can therefore conclude that the analysis of correlations by Markov-switching models confirms the existence of two phases of the stock markets, one characterized by a relatively stable market and one in which market operators behave in a more nervous way.

We can take a step further and augment our model in (6.3) by introducing one lagged value of correlation, whose coefficient is state dependent:

$$\rho_t = \mu_s + \phi_s \rho_{t-1} + \epsilon_t \qquad (6.4)$$

We can fit model (6.4) by specifying the `switch()` option of the `mswitch` command:

```
. tsset newdate
        time variable:  newdate, 02jan1991 to 30dec2013
                delta:  1 day
. mswitch dr corr_SP500_Dax30 if date >= mdy(01,01,2002) & date <= mdy(12,31,2009),
> switch(L.corr_SP500_Dax30)

Performing EM optimization:

Performing gradient-based optimization:

Iteration 0:    log likelihood =  8122.3263
Iteration 1:    log likelihood =  8146.0843
Iteration 2:    log likelihood =  8176.7648
Iteration 3:    log likelihood =  8178.7721
Iteration 4:    log likelihood =  8178.7783
Iteration 5:    log likelihood =  8178.7783

Markov-switching dynamic regression

Sample: 02jan2002 - 30dec2009            No. of obs      =        2,038
Number of states =    2                  AIC             =      -8.0194
Unconditional probabilities: transition  HQIC            =      -8.0123
                                         SBIC            =      -8.0001

Log likelihood = 8178.7783
```

corr_SP500_Dax30	Coef.	Std. Err.	z	P>|z|	[95% Conf. Interval]	
State1						
corr_SP500_Dax30						
L1.	.9907472	.0028298	350.11	0.000	.9852009	.9962935
_cons	-.0034403	.0013795	-2.49	0.013	-.0061441	-.0007366
State2						
corr_SP500_Dax30						
L1.	.9837537	.00132	745.26	0.000	.9811665	.9863409
_cons	.0112318	.0006942	16.18	0.000	.0098711	.0125925
sigma	.0033274	.0000585			.0032148	.003444
p11	.7760997	.0197446			.7350461	.8124155
p21	.0797607	.0074621			.0663063	.0956654

By comparing the results above with the previous table of results, we can see that we now have two additional state-dependent coefficients loading the lagged value of correlations. Moreover, we can see that the information criteria are by far lower in the augmented case, with the AIC moving from -3.47 to -8.02 and the BIC from -3.45 to -8.00. The two coefficients loading the lagged correlations are very close to each other; therefore, we could consider simplifying our model by specifying a constant coefficient for this variable:

$$\rho_t = \mu_s + \phi\rho_{t-1} + \epsilon_t$$

```
. mswitch dr corr_SP500_Dax30 L.corr_SP500_Dax30 if date >= mdy(01,01,2002) &
> date <= mdy(12,31,2009)

Performing EM optimization:

Performing gradient-based optimization:

Iteration 0:    log likelihood =  8137.5737
Iteration 1:    log likelihood =  8164.0564
Iteration 2:    log likelihood =  8175.4104
Iteration 3:    log likelihood =  8175.5395
Iteration 4:    log likelihood =  8175.5396

Markov-switching dynamic regression

Sample: 02jan2002 - 30dec2009            No. of obs      =       2,038
Number of states =    2                  AIC             =     -8.0172
Unconditional probabilities: transition  HQIC            =     -8.0111
                                         SBIC            =     -8.0007

Log likelihood = 8175.5396
```

corr_SP500_Dax30	Coef.	Std. Err.	z	P>|z|	[95% Conf. Interval]	
corr_SP500_Dax30						
corr_SP500_Dax30						
L1.	.9842898	.0013122	750.12	0.000	.981718	.9868617
State1						
_cons	−.0003596	.0006817	−0.53	0.598	−.0016958	.0009766
State2						
_cons	.010938	.00069	15.85	0.000	.0095857	.0122903
sigma	.0033387	.0000587			.0032256	.0034558
p11	.7767789	.019848			.7354866	.8132619
p21	.0786341	.0073748			.0653414	.0943581

The coefficient loading the lagged correlation is equal to 0.9843, a value lying between the two previously estimated values, 0.9907 and 0.9838. The AIC for this model is equal to -8.0172 versus a previous value of -8.0194, and the BIC is equal to -8.0007 versus -8.0001. The two models are therefore almost equivalent.

We now want to focus our attention on Markov-switching AR models. These models are characterized by

$$y_t = \mu_{s_t} + x_t\alpha + z_t\beta_{s_t} + \sum_{i=1}^{p} \phi_{i,s_t}\left(y_{t-i} - \mu_{s_{t-i}} - x_{t-i}\alpha - z_{t-i}\beta_{s_{t-i}}\right) + \varepsilon_{s_t} \qquad (6.5)$$

where y_t is the dependent variable at time t, μ_{s_t} is the state-dependent intercept, x_t are covariates whose coefficients α are state invariant, and z_t are covariates whose coefficients β_{s_t} are state dependent. As in MSDR models, x_t and z_t may contain lags of y_t. ϕ_{i,s_t} are the $i = 1, \dots, p$ AR terms in state s_t. In addition, note that $\mu_{s_{t-i}}$ are the intercepts corresponding to the state that the process was in during the $(t-1)$th period. A similar reading applies to coefficients β.

Representation in (6.5) helps to make clear where the MSAR model distinguishes itself from the MSDR model; that is, here the lagged errors depend on the state the process was in at the previous step.

We now want to fit such a model on the time series of the S&P 500 and DAX 30 correlation:

```
. mswitch ar corr_SP500_Dax30 if date >= mdy(01,01,2002) &
> date <= mdy(12,31,2009), ar(1)

Performing EM optimization:

Performing gradient-based optimization:

numerical derivatives are approximate
nearby values are missing
numerical derivatives are approximate
nearby values are missing
Iteration 0:   log likelihood =   7569.014  (not concave)
could not calculate numerical derivatives -- flat or discontinuous region
> encountered
could not calculate numerical derivatives -- flat or discontinuous region
> encountered
r(430);
```

The model does not converge. Note that Markov-switching models can experience problems in convergence because of the existence of multiple local minimums. When not achieving convergence, we can try to specify an alternative optimization algorithm or to set alternative starting values.

6.2.3   Higher moments contagion

A final approach to contagion analysis recently proposed in the literature by Fry, Martin, and Tang (2010) relies on higher-order conditional moments, which are coskewness and cokurtosis.

Fry, Martin, and Tang (2010) define contagion as a statistically significant increase in the unconditional correlation coefficient, using the Forbes and Rigobon (2002) approach, and they also evaluate significant changes in the higher-order conditional moments of the asset returns, that is, skewness and kurtosis.

By developing an asset pricing model that is an extension of the stochastic discount-factor model from Harvey and Siddique (2000), they derive an explicit expression for pricing risk in terms of risk prices and risk quantities. They define the conditional moments, coskewness, and cokurtosis, and they show that these conditional moments are crucial to the construction of contagion tests. The presence of contagious channels through specifically the coskewness between markets is the focus of their research.

Let us start by defining coskewness as the volatility in one market affecting the mean returns of another market, or alternatively, the mean returns of one market affecting the volatility of another market. Relying on this definition, we can define two tests statistics CS_1 and CS_2, depending on whether the asset market at the source of the crisis is expressed in terms of returns or squared returns:

$$
\mathrm{CS}_1\left(i \to j; r_i^1, r_j^2\right) = \left\{ \frac{\widehat{\psi}_y\left(r_i^1, r_j^2\right) - \widehat{\psi}_x\left(r_i^1, r_j^2\right)}{\sqrt{\left(4\widehat{v}_{y|x_i}^2 + 2\right)/T_y + \left(4\widehat{\rho}_x^2 + 2\right)/T_x}} \right\}^2 \tag{6.6}
$$

$$
\mathrm{CS}_2\left(i \to j; r_i^2, r_j^1\right) = \left\{ \frac{\widehat{\psi}_y\left(r_i^2, r_j^1\right) - \widehat{\psi}_x\left(r_i^2, r_j^1\right)}{\sqrt{\left(4\widehat{v}_{y|x_i}^2 + 2\right)/T_y + \left(4\widehat{\rho}_x^2 + 2\right)/T_x}} \right\}^2 \tag{6.7}
$$

where r_i^1 is the return in the ith market; r_i^2 is the squared return in the ith market; T_x and T_y are sample sizes in the precrisis and the crisis period, respectively; ρ_x and ρ_y are the conditional correlations between markets i and j during the precrisis and the crisis period, respectively; and $v_{y|x_i}$ is the Forbes and Rigobon (2002) adjusted sample correlation. Finally, $\widehat{\psi}_y(r_i^m, r_j^n)$ and $\widehat{\psi}_x(r_i^m, r_j^n)$ take the form

$$
\widehat{\psi}_y\left(r_i^m, r_j^n\right) = \frac{1}{T_y} \sum_{t=1}^{T_y} \left(\frac{y_{i,t} - \widehat{u}_{y_i}}{\widehat{\sigma}_{y_i}}\right)^m \left(\frac{y_{j,t} - \widehat{u}_{y_j}}{\widehat{\sigma}_{y_j}}\right)^n \tag{6.8}
$$

$$
\widehat{\psi}_x\left(r_i^m, r_j^n\right) = \frac{1}{T_x} \sum_{t=1}^{T_x} \left(\frac{x_{i,t} - \widehat{u}_{x_i}}{\widehat{\sigma}_{x_i}}\right)^m \left(\frac{x_{j,t} - \widehat{u}_{x_j}}{\widehat{\sigma}_{x_j}}\right)^n \tag{6.9}
$$

where x denotes the precrisis period and y denotes the crisis period. The null hypothesis is the absence of contagion effects under which $\mathrm{CS}_1(i \to j)$ and $\mathrm{CS}_2(i \to j)$ are distributed as a χ_1^2.

Fry, Martin, and Tang (2010) show that precrisis asset returns display positive returns, low volatility, negative skewness, and negative coskewness. Instead, in periods of financial crisis, returns exhibit lower average returns, high volatility, positive skewness, and positive coskewness.

The distinguishing feature of the coskewness test statistic developed by Fry, Martin, and Tang (2010) is that it does not require specification of the base market from where the contagion originated. In a pair of countries, country i and country j, it is possible to easily test for contagion originating from either country j or country i. In particular, CS_1 evaluates the coskewness measured in terms of returns of country i and squared returns of country j; CS_2 measures the coskewness as squared returns of country i and returns of country j. The contagion is measured mainly through dependence structures between asset markets and changes in dependence structures between asset markets.

Empirical exercise

Starting with the S&P 500 and FTSE 100, we compute the mean and standard deviations for the two series, in the two periods, precrisis and crisis. We define the precrisis period starting from the year 2002 up to the Lehman Brothers collapse of 14 September 2008, and we define the crisis period from 15 September 2008 to the end of 2009.

```
. * SP500
. summarize SP500 if date >= mdy(01,01,2002) & date <= mdy(09,14,2008)
    Variable |        Obs        Mean    Std. Dev.        Min        Max
-------------+--------------------------------------------------------
       SP500 |      1,707      .000027    .0104727   -.0424234    .0557443
. scalar sdSP500_precrisis = r(sd)
. scalar meanSP500_precrisis = r(mean)
. scalar N_precrisis = r(N)
. summarize SP500 if date >= mdy(09,15,2008) & date <= mdy(12,31,2009)
    Variable |        Obs        Mean    Std. Dev.        Min        Max
-------------+--------------------------------------------------------
       SP500 |        331    -.0003769    .0252237   -.0946951    .109572
. scalar sdSP500_crisis = r(sd)
. scalar meanSP500_crisis = r(mean)
. scalar N_crisis = r(N)
. * FTSE100
. summarize FTSE100 if date >= mdy(01,01,2002) & date <= mdy(09,14,2008)
    Variable |        Obs        Mean    Std. Dev.        Min        Max
-------------+--------------------------------------------------------
     FTSE100 |      1,707    .0000252    .0113233   -.0563743    .0590378
. scalar sdFTSE100_precrisis=r(sd)
. scalar meanFTSE100_precrisis=r(mean)
. summarize FTSE100 if date >= mdy(09,15,2008) & date <= mdy(12,31,2009)
    Variable |        Obs        Mean    Std. Dev.        Min        Max
-------------+--------------------------------------------------------
     FTSE100 |        331    -.000016    .0216821   -.0926456    .0938424
. scalar sdFTSE100_crisis = r(sd)
. scalar meanFTSE100_crisis = r(mean)
```

We then compute the Pearson correlation coefficient for each subsample as well as Forbes and Rigobon's (2002) correlation:

```
. * correlation precrisis
. correlate SP500 FTSE100 if date >= mdy(01,01,2002) & date <= mdy(09,14,2008)
(obs=1,707)
             |    SP500  FTSE100
-------------+------------------
       SP500 |   1.0000
     FTSE100 |   0.4587   1.0000

. scalar corr_precrisis = r(rho)
. * crisis Forbes and Rigobon correlation
. correlate SP500 FTSE100 if date >= mdy(09,15,2008) & date <= mdy(12,31,2009)
(obs=331)
             |    SP500  FTSE100
-------------+------------------
       SP500 |   1.0000
     FTSE100 |   0.5966   1.0000
```

```
. scalar corr_crisis=r(rho)

. scalar corrFB_crisis =
> corr_crisis/sqrt(1+(sdSP500_crisis^2/sdSP500_precrisis^2-1)*
> (1-corr_crisis^2))
```

Now we have everything to focus on the test proposed by Fry, Martin, and Tang (2010). In particular, we compute (6.6), (6.7), (6.8), and (6.9).

```
. generate psi_precrisis12 = ((SP500-meanSP500_precrisis)/sdSP500_precrisis)^1*
> ((FTSE100-meanFTSE100_precrisis)/sdFTSE100_precrisis)^2
> if date >= mdy(01,01,2002) & date <= mdy(09,14,2008)
(4,121 missing values generated)

. summarize psi_precrisis12
```

Variable	Obs	Mean	Std. Dev.	Min	Max
psi_precr~12	1,707	.062739	5.72214	-65.58983	87.78837

```
. scalar sum_psi_precrisis12 = r(sum)/r(N)

. generate psi_precrisis21 = ((SP500-meanSP500_precrisis)/sdSP500_precrisis)^2*
> ((FTSE100-meanFTSE100_precrisis)/sdFTSE100_precrisis)^1
> if date >= mdy(01,01,2002) & date <= mdy(09,14,2008)
(4,121 missing values generated)

. summarize psi_precrisis21
```

Variable	Obs	Mean	Std. Dev.	Min	Max
psi_precr~21	1,707	.0280573	5.756753	-58.50124	100.941

```
. scalar sum_psi_precrisis21 = r(sum)/r(N)

. generate psi_crisis12 = ((SP500-meanFTSE100_crisis)/sdSP500_crisis)^1*
> ((FTSE100-meanFTSE100_crisis)/sdFTSE100_crisis)^2
> if date >= mdy(09,15,2008) & date <= mdy(12,31,2009)
(5,497 missing values generated)

. summarize psi_crisis12
```

Variable	Obs	Mean	Std. Dev.	Min	Max
psi_crisis12	331	-.0807318	5.853942	-44.04306	58.23883

```
. scalar sum_psi_crisis12 = r(sum)/r(N)

. generate psi_crisis21 = ((SP500-meanFTSE100_crisis)/sdSP500_crisis)^2*
> ((FTSE100-meanFTSE100_crisis)/sdFTSE100_crisis)^1
> if date >= mdy(09,15,2008) & date <= mdy(12,31,2009)
(5,497 missing values generated)

. summarize psi_crisis21
```

Variable	Obs	Mean	Std. Dev.	Min	Max
psi_crisis21	331	-.2536753	6.03042	-48.26213	69.10927

```
. scalar sum_psi_crisis21 = r(sum)/r(N)

. scalar CS1 = ((sum_psi_crisis12-sum_psi_precrisis12)/
> sqrt((4*corrFB_crisis^2+2)/N_crisis+(4*corr_precrisis^2+2)/N_precrisis))^2

. scalar CS2 = ((sum_psi_crisis21-sum_psi_precrisis21)/
> sqrt((4*corrFB_crisis^2+2)/N_crisis+(4*corr_precrisis^2+2)/N_precrisis))^2

. display "CS1 = " CS1
CS1 = 2.3502755

. display "CS2 = " CS2
CS2 = 9.0628762
```

```
. display "Critical value = " invchi2(1,0.95)
Critical value = 3.8414588
. drop psi_precrisis12 psi_precrisis21 psi_crisis12 psi_crisis21
```

Results denote the presence of contagion from the squared returns of the S&P 500 to the returns of FTSE 100, with the CS_2 test leading to the rejection of the null hypothesis about no contagion. The CS_1 test, evaluating contagion from the returns of the S&P 500 to the squared returns of FTSE 100, instead leads us to not reject the null hypothesis about no contagion.

Now we repeat the exercise for DAX 30 and for Nikkei 225:

```
. * DAX30
. summarize Dax30 if date >= mdy(01,01,2002) & date <= mdy(09,14,2008)
    Variable |        Obs        Mean    Std. Dev.        Min        Max
-------------+--------------------------------------------------------------
       Dax30 |      1,707     .0001108    .0151115  -.0743346   .0755268
. scalar sdDax30_precrisis=r(sd)
. scalar meanDax30_precrisis=r(mean)
. summarize Dax30 if date >= mdy(09,15,2008) & date <= mdy(12,31,2009)
    Variable |        Obs        Mean    Std. Dev.        Min        Max
-------------+--------------------------------------------------------------
       Dax30 |        331    -.0001375   .0236092  -.0733552   .1079747
. scalar sdDax30_crisis = r(sd)
. scalar meanDax30_crisis = r(mean)
. * correlation precrisis
. correlate SP500 Dax30 if date >= mdy(01,01,2002) & date <= mdy(09,14,2008)
(obs=1,707)
             |    SP500      Dax30
-------------+------------------
       SP500 |   1.0000
       Dax30 |   0.5812     1.0000
. scalar corr_precrisis = r(rho)
. * crisis Forbes and Rigobon correlation
. correlate SP500 Dax30 if date >= mdy(09,15,2008) & date <= mdy(12,31,2009)
(obs=331)
             |    SP500      Dax30
-------------+------------------
       SP500 |   1.0000
       Dax30 |   0.6823     1.0000
. scalar corr_crisis=r(rho)
. scalar corrFB_crisis =
> corr_crisis/sqrt(1+(sdSP500_crisis^2/sdSP500_precrisis^2-1)*
> (1-corr_crisis^2))
. generate psi_precrisis12 = ((SP500-meanSP500_precrisis)/sdSP500_precrisis)^1*
> ((Dax30-meanDax30_precrisis)/sdDax30_precrisis)^2 if date >= mdy(01,01,2002) &
> date <= mdy(09,14,2008)
(4,121 missing values generated)
```

```
. summarize psi_precrisis12

    Variable |       Obs        Mean    Std. Dev.        Min         Max
-------------+--------------------------------------------------------------
 psi_precr~12 |     1,707    .1575339    6.758374   -64.27853    125.1884
. scalar sum_psi_precrisis12 = r(sum)/r(N)

. generate psi_precrisis21 = ((SP500-meanSP500_precrisis)/sdSP500_precrisis)^2*
> ((Dax30-meanDax30_precrisis)/sdDax30_precrisis)^1 if date >= mdy(01,01,2002) &
> date <= mdy(09,14,2008)
(4,121 missing values generated)

. summarize psi_precrisis21

    Variable |       Obs        Mean    Std. Dev.        Min         Max
-------------+--------------------------------------------------------------
 psi_precr~21 |     1,707    .1056625    6.417097   -65.42844    126.0845
. scalar sum_psi_precrisis21 = r(sum)/r(N)

. generate psi_crisis12 = ((SP500-meanDax30_crisis)/sdSP500_crisis)^1*
> ((Dax30-meanDax30_crisis)/sdDax30_crisis)^2 if date >= mdy(09,15,2008) &
> date <= mdy(12,31,2009)
(5,497 missing values generated)

. summarize psi_crisis12

    Variable |       Obs        Mean    Std. Dev.        Min         Max
-------------+--------------------------------------------------------------
  psi_crisis12 |       331    .3042933    7.919537   -30.18322    91.20583
. scalar sum_psi_crisis12 = r(sum)/r(N)

. generate psi_crisis21 = ((SP500-meanDax30_crisis)/sdSP500_crisis)^2*
> ((Dax30-meanDax30_crisis)/sdDax30_crisis)^1 if date >= mdy(09,15,2008) &
> date <= mdy(12,31,2009)
(5,497 missing values generated)

. summarize psi_crisis21

    Variable |       Obs        Mean    Std. Dev.        Min         Max
-------------+--------------------------------------------------------------
  psi_crisis21 |       331    .0119648    7.552359   -39.87618    86.62904
. scalar sum_psi_crisis21 = r(sum)/r(N)

. scalar CS1 =
((sum_psi_crisis12-sum_psi_precrisis12)/sqrt((4*corrFB_crisis^2+2)/N_crisis+
> (4*corr_precrisis^2+2)/N_precrisis))^2

. scalar CS2 =
> ((sum_psi_crisis21-sum_psi_precrisis21)/sqrt((4*corrFB_crisis^2+2)/N_crisis+
> (4*corr_precrisis^2+2)/N_precrisis))^2

. display "CS1 = " CS1
CS1 = 2.2474568

. display "CS2 = " CS2
CS2 = .91608828

. display "Critical value = " invchi2(1,0.95)
Critical value = 3.8414588

. drop psi_precrisis12 psi_precrisis21 psi_crisis12 psi_crisis21

. * Nikkei225
. summarize Nikkei225 if date >= mdy(01,01,2002) & date <= mdy(09,14,2008)

    Variable |       Obs        Mean    Std. Dev.        Min         Max
-------------+--------------------------------------------------------------
   Nikkei225 |     1,707    .0000235    .0131127   -.0581569    .0573523
. scalar sdNikkei225_precrisis=r(sd)
```

```
. scalar meanNikkei225_precrisis=r(mean)
. summarize Nikkei225 if date >= mdy(09,15,2008) & date <= mdy(12,31,2009)
```

Variable	Obs	Mean	Std. Dev.	Min	Max
Nikkei225	331	-.0005114	.0262008	-.1211103	.1323458

```
. scalar sdNikkei225_crisis = r(sd)
. scalar meanNikkei225_crisis = r(mean)
. * correlation precrisis
. correlate SP500 Nikkei225 if date >= mdy(01,01,2002) & date <= mdy(09,14,2008)
(obs=1,707)
```

	SP500 Nikk~225
SP500	1.0000
Nikkei225	0.1085 1.0000

```
. scalar corr_precrisis = r(rho)
. * crisis Forbes and Rigobon correlation
. correlate SP500 Nikkei225 if date >= mdy(09,15,2008) & date <= mdy(12,31,2009)
(obs=331)
```

	SP500 Nikk~225
SP500	1.0000
Nikkei225	0.1190 1.0000

```
. scalar corr_crisis=r(rho)
. scalar corrFB_crisis =
> corr_crisis/sqrt(1+(sdSP500_crisis^2/sdSP500_precrisis^2-1)*(1-corr_crisis^2))
. generate psi_precrisis12 =
> ((SP500-meanSP500_precrisis)/sdSP500_precrisis)^1*
> ((Nikkei225-meanNikkei225_precrisis)/sdNikkei225_precrisis)^2
> if date >= mdy(01,01,2002) & date <= mdy(09,14,2008)
(4,121 missing values generated)
. summarize psi_precrisis12
```

Variable	Obs	Mean	Std. Dev.	Min	Max
psi_precr~12	1,707	-.0365796	2.748509	-24.95494	41.79678

```
. scalar sum_psi_precrisis12 = r(sum)/r(N)
. generate psi_precrisis21 =
> ((SP500-meanSP500_precrisis)/sdSP500_precrisis)^2*
> ((Nikkei225-meanNikkei225_precrisis)/sdNikkei225_precrisis)^1
> if date >= mdy(01,01,2002) & date <= mdy(09,14,2008)
(4,121 missing values generated)
. summarize psi_precrisis21
```

Variable	Obs	Mean	Std. Dev.	Min	Max
psi_precr~21	1,707	-.0931086	3.288445	-57.41635	52.51463

```
. scalar sum_psi_precrisis21 = r(sum)/r(N)
. generate psi_crisis12 =
> ((SP500-meanNikkei225_crisis)/sdSP500_crisis)^1*
> ((Nikkei225-meanNikkei225_crisis)/sdNikkei225_crisis)^2
> if date >= mdy(09,15,2008) & date <= mdy(12,31,2009)
(5,497 missing values generated)
```

```
. summarize psi_crisis12
    Variable |        Obs        Mean    Std. Dev.        Min        Max
-------------+--------------------------------------------------------------
 psi_crisis12 |        331    .0493704    3.502988   -20.13489   35.39534
. scalar sum_psi_crisis12 = r(sum)/r(N)

. generate psi_crisis21 =
> ((SP500-meanNikkei225_crisis)/sdSP500_crisis)^2*
> ((Nikkei225-meanNikkei225_crisis)/sdNikkei225_crisis)^1
> if date >= mdy(09,15,2008) & date <= mdy(12,31,2009)
(5,497 missing values generated)

. summarize psi_crisis21
    Variable |        Obs        Mean    Std. Dev.        Min        Max
-------------+--------------------------------------------------------------
 psi_crisis21 |        331   -.1392609    3.267328   -20.23126   39.83128
. scalar sum_psi_crisis21 = r(sum)/r(N)

. scalar CS1 =
> ((sum_psi_crisis12-sum_psi_precrisis12)/sqrt((4*corrFB_crisis^2+2)/
> N_crisis+(4*corr_precrisis^2+2)/N_precrisis))^2

. scalar CS2 =
> ((sum_psi_crisis21-sum_psi_precrisis21)/sqrt((4*corrFB_crisis^2+2)/
> N_crisis+(4*corr_precrisis^2+2)/N_precrisis))^2

. display "CS1 = " CS1
CS1 = 1.0159544

. display "CS2 = " CS2
CS2 = .29293382

. display "Critical value = " invchi2(1,0.95)
Critical value = 3.8414588
```

The results show the absence of contagion for both pairs of countries analyzed and for both the CS1 and the CS2 test statistics.

Glossary of acronyms

AC	autocorrelation
ACF	autocorrelation function
ADF	augmented Dickey–Fuller
AGARCH	asymmetric generalized autoregressive conditional heteroskedasticity
AGDCC	asymmetric generalized dynamic conditional correlation
AIC	Akaike information criterion
APARCH	asymmetric power ARCH
AR	autoregressive
ARCH	autoregressive conditional heteroskedasticity
ARCH-M	ARCH-in-mean
ARIMA	autoregressive integrated moving average
ARMA	autoregressive moving average
ARMAX	augmented ARMA
BEKK	Baba–Engle–Kraft–Kroner
BFGS	Broyden–Fletcher–Goldfarb–Shanno
BHHH	Berndt–Hall–Hall–Hausman
BIC	Bayesian information criterion
cc	conditional coverage (backtesting procedure)
CCC	constant conditional correlation
c.d.f.	cumulative distribution function
DCC	dynamic conditional correlation
DCCE	DCC (Engle)
DCC-MIDAS	DCC mixed data sampling
DCCT	DCC (Tse and Tsui)
DF-GLS	Dickey–Fuller generalized least squares
DFP	Davidon–Fletcher–Powell
dvech	diagonal vech
EACD	exponential autoregressive conditional duration
ES	expected shortfall
EVT	extreme value theory
F-GARCH	GARCH relying on factor decomposition
factor GARCH	factor generalized autoregressive conditional heteroskedasticity
GARCH	generalized autoregressive conditional heteroskedasticity
GARCH-M	GARCH-in-mean
GED	generalized error distribution

GJR	Glosten–Jagannathan–Runkle
GLS	generalized least squares
GMM	generalized method of moments
IGARCH	integrated GARCH
i.i.d.	independent and identically distributed
ind	independence (backtesting procedure)
LM	Lagrange multiplier
LR	likelihood ratio
MA	moving average
MAIC	modified Akaike information criterion
MGARCH	multivariate GARCH
MIDAS	mixed data sampling
ML	maximum likelihood
MSAR	Markov-switching autoregression
MSDR	Markov-switching dynamic regression
MSE	mean squared error
NGARCH	nonlinear GARCH
NGARCHK	nonlinear GARCH with one shift
NIC	news impact curve
NR	Newton–Raphson
PAC	partial autocorrelation
PACF	partial autocorrelation function
PARCH	power ARCH
p.d.f.	probability density function
P&L	profit and loss
PP	Phillips and Perron
QLIKE	quasilikelihood
QML	quasimaximum likelihood
SAARCH	simple asymmetric ARCH
TGARCH	threshold GARCH
uc	unconditional coverage (backtesting procedure)
VaR	value-at-risk
VAR	vector autoregressive
VCE	variance–covariance estimate

References

Baba, Y., R. F. Engle, D. F. Kraft, and K. F. Kroner. 1991. Multivariate simultaneous generalized ARCH. Discussion Paper No. 89, Department of Economics, University of California–San Diego.

Bae, K., G. A. Karolyi, and R. M. Stulz. 2003. A new approach to measuring financial contagion. *Review of Financial Studies* 16: 717–763.

Bauwens, L., C. Hafter, and S. Laurent. 2011. Volatility models. Technical Report 2011/58, CORE discussion paper, Center for Operations Research and Econometrics. http://uclouvain.be/cps/ucl/doc/core/documents/coredp2011_58web.pdf.

———. 2012. *Handbook of Volatility Models and Their Applications*. Hoboken, NJ: Wiley.

Bauwens, L., S. Laurent, and J. V. K. Rombouts. 2006. Multivariate GARCH models: A survey. *Journal of Applied Econometrics* 21: 79–109.

Berkowitz, J., and J. O'Brien. 2002. How accurate are value-at-risk models at commercial banks? *Journal of Finance* 57: 1093–1111.

Black, F. 1976. The pricing of commodity contracts. *Journal of Financial Economics* 3: 167–179.

Bollerslev, T. 1986. Generalized autoregressive conditional heteroskedasticity. *Journal of Econometrics* 31: 307–327.

———. 1990. Modelling the coherence in short-run nominal exchange rates: A multivariate generalized ARCH model. *Review of Economics and Statistics* 72: 498–505.

———. 2001. Financial econometrics: Past developments and future challenges. *Journal of Econometrics* 100: 41–51.

———. 2010. Glossary to ARCH (GARCH*). In *Volatility and Time Series Econometrics: Essays in Honor of Robert Engle*, ed. T. Bollerslev, J. R. Russell, and M. W. Watson, chap. 8. Oxford: Oxford University Press.

Bollerslev, T., R. F. Engle, and D. B. Nelson. 1994. ARCH models. In Vol. 4 of *Handbook of Econometrics*, ed. R. F. Engle and D. L. McFadden, chap. 49. Amsterdam: Elsevier.

Bollerslev, T., R. F. Engle, and J. M. Wooldridge. 1988. A capital asset pricing model with time-varying covariances. *Journal of Political Economy* 96: 116–131.

Bollerslev, T., and J. M. Wooldridge. 1992. Quasi-maximum likelihood estimation and inference in dynamic models with time-varying covariances. *Econometric Reviews* 11: 143–172.

Bozdogan, H. 1987. Model selection and Akaike's information criterion (AIC): The general theory and its analytical extensions. *Psychometrika* 52: 345–370.

Cappiello, L., R. F. Engle, and K. Sheppard. 2006. Asymmetric dynamics in the correlations of global equity and bond returns. *Journal of Financial Econometrics* 4: 537–572.

Chan, K. 1993. Imperfect information and cross-autocorrelation among stock prices. *Journal of Finance* 4: 1211–1230.

Chiang, T. C., B. N. Jeon, and H. Li. 2007. Dynamic correlation analysis of financial contagion: Evidence from Asian markets. *Journal of International Money and Finance* 26: 1206–1228.

Christoffersen, P. F. 1998. Evaluating interval forecasts. *International Economic Review* 39: 841–862.

Christoffersen, P. F., and D. Pelletier. 2004. Backtesting value-at-risk: A duration-based approach. *Journal of Financial Econometrics* 2: 84–108.

Colacito, R., R. F. Engle, and E. Ghysels. 2011. A component model for dynamic correlations. *Journal of Econometrics* 164: 45–59.

D'Agostino, R. B., A. J. Belanger, and R. B. D'Agostino, Jr. 1990. A suggestion for using powerful and informative tests of normality. *American Statistician* 44: 316–321.

Dickey, D. A., and W. A. Fuller. 1979. Distribution of the estimators for autoregressive time series with a unit root. *Journal of the American Statistical Association* 74: 427–431.

Ding, Z., C. W. J. Granger, and R. F. Engle. 1993. A long memory property of stock market returns and a new model. *Journal of Empirical Finance* 1: 83–106.

Dufour, J. 2006. Monte Carlo tests with nuisance parameters: A general approach to finite-sample inference and nonstandard asymptotics. *Journal of Econometrics* 133: 443–477.

Elliott, G., T. J. Rothenberg, and J. H. Stock. 1996. Efficient tests for an autoregressive unit root. *Econometrica* 64: 813–836.

Engle, R. F. 1982. Autoregressive conditional heteroskedasticity with estimates of the variance of United Kingdom inflation. *Econometrica* 50: 987–1007.

———. 1990. Discussion: Stock volatility and the crash of '87. *Review of Financial Studies* 3: 103–106.

———. 2001. Financial econometrics—a new discipline with new methods. *Journals of Econometrics* 100: 53–56.

———. 2002. Dynamic conditional correlation: A simple class of multivariate generalized autoregressive conditional heteroskedasticity models. *Journal of Business and Economic Statistics* 20: 339–350.

Engle, R. F., and G. Gonzalez-Rivera. 1991. Semiparametric ARCH models. *Journal of Business and Economic Statistics* 9: 345–359.

Engle, R. F., and C. W. J. Granger. 2003. Time-series econometrics: Cointegration and autoregressive conditional heteroskedasticity. Technical report, Advanced information on the Bank of Sweden Prize in Economic Sciences in Memory of Alfred Nobel, Stockholm, Sweden. http://www.nobelprize.org/nobel_prizes/economic-sciences/laureates/2003/advanced-economicsciences2003.pdf.

Engle, R. F., and V. K. Ng. 1993. Measuring and testing the impact of news on volatility. *Journal of Finance* 48: 1749–1778.

Engle, R. F., and J. R. Russell. 1998. Autoregressive conditional duration: A new model for irregularly spaced transaction data. *Econometrica* 66: 1127–1162.

Engle, R. F., and K. Sheppard. 2001. Theoretical and empirical properties of dynamic conditional correlation multivariate GARCH. Mimeo, University of California–San Diego. https://archive.nyu.edu/bitstream/2451/26570/2/FIN-01-027.pdf.

Fisher, R. A. 1915. Frequency distribution of the values of the correlation coefficient in samples from an indefinitely large population. *Biometrika* 10: 507–521.

Forbes, K. J., and R. Rigobon. 2002. No contagion, only interdependence: Measuring stock market comovements. *Journal of Finance* 57: 2223–2261.

Fry, R., V. L. Martin, and C. Tang. 2010. A new class of tests of contagion with applications. *Journal of Business and Economic Statistics* 28: 423–437.

Ghysels, E., P. Santa-Clara, and R. Valkanov. 2004. The MIDAS touch: Mixed data sampling regression models. Technical report, Discussion Paper, University of California–Los Angeles. http://rady.ucsd.edu/faculty/directory/valkanov/pub/docs/midas-touch.pdf.

———. 2005. There is a risk-return trade-off after all. *Journal of Financial Economics* 76: 509–548.

———. 2006. Predicting volatility: Getting the most out of return data sampled at different frequencies. *Journal of Econometrics* 131: 59–95.

Ghysels, E., A. Sinko, and R. Valkanov. 2007. MIDAS regressions: Further results and new directions. *Econometric Reviews* 26: 53–90.

Glosten, L. R., R. Jagannathan, and D. E. Runkle. 1993. On the relation between the expected value and the volatility of the nominal excess return on stocks. *Journal of Finance* 48: 1779–1801.

Goldfeld, S. M., and R. E. Quandt. 1973. A Markov model for switching regressions. *Journal of Econometrics* 1: 3–15.

Hamilton, J. D. 1989. A new approach to the economic analysis of nonstationary time series and the business cycle. *Econometrica* 57: 357–384.

Hansson, B., and P. Hordahl. 1998. Testing the conditional CAPM using multivariate GARCH-M. *Applied Financial Economics* 8: 377–388.

Harvey, C. R., and A. Siddique. 2000. Conditional skewness in asset pricing tests. *Journal of Finance* 55: 1263–1295.

Higgins, M. L., and A. K. Bera. 1992. A class of nonlinear ARCH models. *International Economic Review* 33: 137–158.

Hurlin, C., and S. Tokpavi. 2006. Backtesting value-at-risk accuracy: A simple new test. *Journal of Risk* 9: 19–37.

Kupiec, H. 1995. Techniques for verifying the accuracy of risk measurement models. No. 95-24, Finance and Economics Discussion Series, Board of Governors of the Federal Reserve System (U.S.).

Leccadito, A., S. Boffelli, and G. Urga. 2014. Evaluating the accuracy of value-at-risk forecasts: New multilevel tests. *International Journal of Forecasting* 30: 206–216.

Ljung, G. M., and G. E. P. Box. 1978. On a measure of lack of fit in time series models. *Biometrika* 65: 297–303.

Newey, W. K., and K. D. West. 1987. A simple, positive semi-definite, heteroskedasticity and autocorrelation consistent covariance matrix. *Econometrica* 55: 703–708.

Patton, A. J. 2011. Volatility forecast comparison using imperfect volatility proxies. *Journal of Econometrics* 160: 246–256.

Perignon, C., and D. Smith. 2008. A new approach to comparing VaR estimation methods. *Journal of Derivatives* 16: 54–66.

Phillips, P. C. B., and P. Perron. 1988. Testing for a unit root in time series regression. *Biometrika* 75: 335–346.

Quandt, R. E. 1972. A new approach to estimating switching regressions. *Journal of the American Statistical Association* 67: 306–310.

Silvennoinen, A., and T. Teräsvirta. 2009. Multivariate GARCH models. In *Handbook of Financial Time Series*, ed. T. G. Andersen, R. A. Davis, J.-P. Kreiß, and T. Mikosch, 201–229. New York: Springer.

Teräsvirta, T. 2009. An introduction to univariate GARCH models. In *Handbook of Financial Time Series*, ed. T. G. Andersen, R. A. Davis, J.-P. Kreiß, and T. Mikosch, 17–41. New York: Springer.

Tse, Y. K. 2000. A test for constant correlations in a multivariate GARCH model. *Journal of Econometrics* 98: 107–127.

Tse, Y. K., and A. K. C. Tsui. 2002. A multivariate generalized autoregressive conditional heteroscedasticity model with time-varying correlations. *Journal of Business and Economic Statistics* 20: 351–362.

Weiss, A. A. 1986. Asymptotic theory for ARCH models: Estimation and testing. *Econometric Theory* 2: 107–131.

Zakoian, J. M. 1994. Threshold heteroskedastic models. *Journal of Economic Dynamics and Control* 18: 931–955.

Zivot, E. 2009. Practical issues in the analysis of univariate GARCH models. In *Handbook of Financial Time Series*, ed. T. G. Andersen, R. A. Davis, J.-P. Kreiß, and T. Mikosch, 113–155. New York: Springer.

Author index

Subject index